'Experienced from many years of teaching and supervising in higher education, I can say that Alvehus' elegant guidance and introduction to the mysteries of writing a good student thesis, is extremely useful to students as well as teachers and supervisors.'

Mikael Askander, *Associate Professor in Intermedial Studies, Lund University (Sweden)*

'In the vast realm of textbooks on qualitative methods, Alvehus offers something different. The book reads, almost, as a kind of Socratic dialogue between teacher and learner, where complex ideas are succinctly described in ways that attends to understanding and application, and providing readers with the confidence to engage in further and deeper reading. As a kind of myth-buster, it provides an accessible, conversational and imaginative response to the common stumbling blocks faced by students.'

Professor Justin Waring, *Professor of Medical Sociology, University of Birmingham*

'Doing qualitative research and writing it up is one of the most difficult tasks that our students face. And yet, mastering this task engages students in one of the most important learning and development processes. Supporting students throughout this journey is not an easy feat and many textbooks available on the matter use complicated jargon that raises more questions than it answers. This handbook does the opposite. It starts from a student's perspective and answers their questions directly and in an accessible way. It will soon become the number one resource for students and supervisors alike.'

Doris Schedlitzki, *Professor of Organisational Leadership and Head of Research, Guildhall School of Business and Law, London Metropolitan University*

CRAFTING YOUR THESIS

At the beginning of writing a thesis, many questions arise, for example:

- How do I know that I have formulated a relevant research problem?
- Have I chosen the right empirical method?
- Are interviews or observations appropriate?
- How should I structure my text to get my point across in the best way?
- What exactly is a theory?
- How can the quality of my work be assessed?

Crafting Your Thesis is a broad and accessible handbook in qualitative methods that gives you clear and concise answers to these questions – and many more. The book can be used both in introductory university courses, where you as a student encounter questions of method for perhaps the first time, and right up to Master's thesis level, where it gives a quick overview of different available qualitative methods and highlights questions that must be dealt with when crafting the thesis.

Johan Alvehus is professor at the Department of Service Studies, Lund University.

CRAFTING YOUR THESIS

Making Use of Qualitative Approaches

Johan Alvehus

Routledge
Taylor & Francis Group

LONDON AND NEW YORK

Cover image: Getty Images © MirageC

First published 2025
by Routledge
4 Park Square, Milton Park, Abingdon, Oxon OX14 4RN

and by Routledge
605 Third Avenue, New York, NY 10158

Routledge is an imprint of the Taylor & Francis Group, an informa business

British Library Cataloguing-in-Publication Data
A catalogue record for this book is available from the British Library

ISBN: 978-1-032-81156-7 (hbk)
ISBN: 978-1-032-79958-2 (pbk)
ISBN: 978-1-003-49838-4 (ebk)

DOI: 10.4324/9781003498384

Typeset in Times New Roman
by MPS Limited, Dehradun

CONTENTS

PREFACE TO THE ENGLISH EDITION

The idea for the book you are holding in your hand, or seeing on your screen, was conceived during a semester when I was supervising an unusually large number of undergraduate theses. During the supervision process, I began to reflect on the fact that certain questions tended to recur, and in a rare act of proactiveness, I put together a collection of answers to those questions. That collection eventually became the backbone of this book, and it is the reason why the chapter headings in this book look the way they do; several of them are exactly the questions I received from students during that semester.

The ambition of this book has always been to provide a simple and easy-to-use introduction to a topic that many students struggle with. Students' questions, as just mentioned, are very much at the center. The ambition seems to have worked. At the time of writing, the Swedish edition of the book has been cited over four thousand times, according to Google Scholar. I can't imagine a better reception. I have also received encouraging comments from students who have said that 'finally I understand how an academic text works'. There is clearly a need for a relatively straightforward approach to the questions posed in the book, and I am of course happy that these particular answers have been able to help many. With that in mind, I have tried to maintain the basic approach of the book in this English edition.

In a way, this is the fourth edition of the book. The book was first published in 2013 in Swedish, and it was thoroughly revised for its third edition in 2023. This English translation is a slightly revised version of the third edition, adapted for an international audience. The latter means that I have tried to, where possible and appropriate, make references to literature available in English. As this book is by design very short, it should serve as an introduction to other literature on methods, and therefore relevant references are key.

While working on the different editions of this book, I have had the help of a number of friends and colleagues who have read and commented. Many thanks to Eerika Saaristo, Fredrik Nilsson, and Nils-Göran Olve for their input. For the third edition, I had the help of students who read and commented on earlier versions. Special thanks to Sam Alvehus, Alva Rosager, and Erik Willman for reading and commenting. Karin Alvehus has, as always, been the critical reader par excellence. Also, thank you, Adam Woods at Routledge, for taking on this project and supporting me throughout.

1

INTRODUCTION

SO, YOU'RE WRITING A THESIS? Maybe you will spend a few weeks on it, maybe a few months. Your ambition is of course to write as good a thesis as possible. But how do you do that? And what role does method play in this?

Writing a thesis is one of the most difficult (and fun!) things you'll face in higher education. A thesis is a way of practicing and examining many basic academic skills: The ability to formulate sharp and relevant problems; the ability to reason clearly, systematically, and abstractly; the ability to carry out insightful analysis and come up with solutions to problems. This means that several different skills come together in the thesis, and when the complex interplay between them works, the result is an interesting and readable text. Thesis writing is a difficult art form to master – but it's not impossible! It is my belief that everyone can learn to write a good thesis.

Theses vary in length, from shorter theses written within a course to projects running for several months. They can be of different types. Sometimes they are short papers with a clear theoretical angle and with a research-related contribution, while others aim to make a more practical contribution. If we go outside the walls of the university, they may be investigations or inquiries, or, for example, texts for debate about society. Though there are many different types of theses in terms of both scope and focus, there are issues that are central to them all. And such issues are the concern of this book.

Thus, when I refer to 'thesis' in this book, it is a shorthand. The book refers as much to essays, reports, papers, and similar. As you get to know different genres, you will learn how different parts of the book are relevant in different contexts.

Purpose and target audience of the book

This book is targeted at those who are writing a thesis based on qualitative methods. The aim of the book is to introduce, in a clear and concise way, a series of questions that you as a beginner might ask yourself when writing your thesis. The book is structured around a number of questions that all touch on key elements of thesis writing. The different chapters address both key scientific concepts (such as 'theory' and 'data') and more

DOI: 10.4324/9781003498384-1

practical issues that you as an author will face (such as how to combine different methods, how to do interviews, and how to edit your text).

The book is intended to be used at different levels of education. It can be used during an introductory course, where students are confronted with methodological conundrums for the first time, all the way up to Master's level, where it provides a quick overview of the different methods available and points to issues that can be discussed in seminars and in the supervision process. At higher educational levels, this book should be used in conjunction with other literature and can then provide a platform for discussing the overall design of a thesis and how the different parts relate to each other. Methodological literature is often either very comprehensive or focused on specific sub-problems, and books that provide a quick overview are rare. That is where this book fills an important role.

The ambition of the book: A handbook

Writing method is not easy. Faced with their first such task, many are left feeling helpless. It is not always easy to see how what is written in the methodology books, which are often thick and may even contradict each other, can be translated into the text they are about to write. The methodology part of the thesis therefore often becomes a kind of compliant presentation of answers to questions one expects to receive from the supervisor or from opponents at a seminar. The methodology chapter frequently begins with a discussion of 'positivism versus hermeneutics', 'induction versus deduction', 'quantitative versus qualitative', or some other central, abstract, and extremely exhausted (and boring) question from the philosophy of science. This is followed by a description of the method(s) used and then an account of what was actually done (for example, how many observations were made or how long the interviews were). Finally, there is often a 'methodological critique' which discusses the credibility of the thesis.

This way of discussing and understanding the role of the method in the thesis misses its target. The methods question risks becoming something that is only relevant in the methods chapter. But method in interpretive qualitative research is not something that can be easily reduced to a chapter. Instead, it is a way of reasoning scientifically that is also reflected in other parts of the thesis and that is directly related to the aim, the formulation of the research problem, and the conclusions. Method is not just a set of techniques for capturing 'reality out there' but also a matter of style and of building an argument.

Treating method in this more comprehensive way may seem a bit mysterious. It is not easy to see how to make the line of reasoning and structure of the text really work in such a way that the method permeates the text. This book is a handbook that seeks to provide guidance on just that. The idea is that you, as a thesis writer, will be able to gain support from this book to get a broader understanding of the role of the method in the thesis, but also that you will be able to get concrete and tangible tips and ideas on how to make this happen in practice: In *your* thesis, in *your* writing process.

I also hope that this book will help to broaden the way in which method is treated more generally. Method is an important part of the scientific craft, but it is also of central importance for you in work life and as a citizen.

A mini-encyclopedia

Each chapter in this book addresses a specific question that you, as a thesis writer, might ask yourself. The book serves as a reference to which you can turn for a quick overview of a specific question and the considerations it entails. The questions are divided into four groups. The first, *Method*, deals with questions of a more general nature, relating to method and methodological thinking in a broader sense. It discusses what method is, what it is for, and concepts such as 'qualitative', 'empirical', and 'theory'.

The second part of the book, *Writing*, deals with general issues for writing the thesis. It begins with a discussion of the structure of the thesis and goes on to address the question of what comprises a research problem. It then discusses language and style, text editing, and references, topics that can be tricky enough. The final chapter looks at the writing process, i.e. how you can work to create a well-written thesis.

The third part of the book, *Fieldwork*, deals with the considerations you will make during the creation of your empirical material. First, it discusses the importance of sampling and the various methods for doing so. This is followed by sections on different methods: Case studies, interviews, focus groups, observations, and ethnography. Of course, these should not be seen as a complete list of all available qualitative methods. There is a plethora of other empirical and analytical methods that are not covered in detail here. The methods not included in the book are of course no less appropriate or less important. The selection in this book is based on what I have found is commonly used at Bachelor's and Master's level. For those using methods not covered in this book, the other chapters still have value.

Finally, there are three chapters grouped under the heading of *At the desk*. They deal with analysis, critique, and quality – parts of the work that you usually deal with mainly at the desk and on the computer desktop, although they are all present in an entire writing process. They also have implications for life in general, for life outside the confines of a thesis. They are, however, very much present in the final stages of completing the thesis, when everything has to fall into place. That is the reason for ending the book with these issues, but again, they are certainly not confined to the final stages of thesis writing only.

The encyclopedia character of the book means that the different chapters do not directly build on each other. Anyone who is more curious about fieldwork would do well to start in these chapters. In the most obvious places, there are references to other chapters in the book to facilitate this type of reading.

The book is deliberately brief and seeks to provide not the details but the overview. Sometimes this leads to nuances being lost, but the details and nuances on the other hand come through in the other methodological literature that is still needed when writing a more comprehensive work. The book takes an overview approach in that it shows how the different parts of the thesis are connected from a methodological point of view. In addition to this, you will of course need to take into account the methods, theories, and current topics of the particular research field in which your thesis is positioned.

PART 1

Method

2

WHY METHOD?

A question one might ask – and that many do ask – is what on Earth does methodology have to do with 'reality out there'? Isn't methodology just something you do at university? For many, a Bachelor's or Master's thesis is the only time in their lives that they write anything substantial about method. However, that's not the same as it being the only time they are confronted with method. On the contrary: Method, in the broader sense that this book represents, is something we encounter all the time.

Methodological skills in everyday life

Almost daily, we are confronted with the findings of various studies, with more or less confident assertions about how things are or work. Perhaps the most obvious are polls that are conducted and reported, particularly at election time, some of which are obviously flawed (e.g. some web-based polls), while others are presented with a clear indication of both the question and the statistical margin of error. In many contexts in work life, we are confronted with workplace surveys, market research, and reports on various issues. Sometimes we see studies where they claim to have interviewed several hundred participants – where they are in fact telephone surveys. Sometimes we see a study based on 30 participants that is taken to be representative for a population in general. Often, more data seems to be equated with better – but is that always the case? We are asked to make decisions on everything from what to eat to how to raise our children based on the results of various types of studies, broadcasted on tabloid front pages and in our social media news feeds. In all of these cases, and many more, it is of advantage to know something about how different methods work and what kind of conclusions can be drawn from them. Only then can one make sensible use of and be critical of the results of the surveys with which one is confronted.

But, as this book will argue, method is about more than that. Method is not just about how 'data about reality' is collected, processed, and presented, but perhaps more importantly about how a particular image of reality is produced: How the problem is presented, what questions are asked, what concepts are used, and how the underlying

DOI: 10.4324/9781003498384-3

arguments for suggestions about solutions are constructed. Take for example education, which is constantly in the public debate. There are lots of claims about what is not working, why it is not working, and what should be done about it. In order to be critical of that debate, it is essential to know something about how not only studies are conducted, but above all how problematizations are constructed and how arguments are built up: Why is question A asked and not question B or C? Why are results X or Y not asked for, while Z is given a prominent role? And how does one arrive at the conclusion that M is better than N? Methodological skills are also about understanding how questions are asked and how arguments are constructed, skills that go far beyond individual methods and techniques.

Methodology in higher education

In higher education, you are confronted with methodological issues in several ways. Partly in the form of methods courses, where you may practice conducting interviews or observations, and partly through the texts you read, which are of course based on methodological considerations. This means that methodological knowledge is not only relevant when you write your own texts – theses, reports, or papers – but also when you read the texts of others. If you understand method as something that permeates a text, it becomes something that should always be present in your reading. Although this book is written primarily from the point of view that you are about to write a thesis yourself, it contains many questions that should be asked of other texts that you encounter in your studies. Using the U-model (see Chapter 6) as a starting point is a good way to analyze how a text holds together and to analyze its argumentation. When reading a study and thinking about its contribution, you should consider whether the empirical material really answers the question posed, and moreover, the theoretical contribution of the study must be evaluated against the theoretical ambitions initially outlined. The CARS model (see Chapter 7) can be used to analyze whether a problem statement is relevant or not and whether it should be tweaked in some way.

Methodological knowledge thus contributes to a type of reading of scientific texts that more carefully analyzes and criticizes what the text tries to convey. One reason why methods courses often appear early in higher education is to encourage a more active, reflective, and critical search for knowledge. There is an important point in designing a curriculum in this way, even if it may seem that methodology is an alien subject that has mostly to do with research. But, as mentioned, methodology can be understood from a broader perspective.

A critical approach

To be critical is almost always presented as a key part of any scientific effort. Phrases such as 'engage critically', 'critically reflect', and 'critically assess' keep recurring.

However, the question is what this 'critical' actually means. Is it to be skeptical in general and dismiss anything that goes against what you thought you already knew? Or does it mean something more? I will discuss the concept of critique in detail later in the book, but for now, I want to emphasize role methodology has in critical thinking. To be able to critically assess a knowledge claim requires both a knowledge of the field in relation to which the claim is made and an understanding of how knowledge claims are made in that particular field of study – that is, method. Critique, in such contexts, is not primarily based on general skepticism or on holding a different opinion from the one expressed. '[C]ritique is expected to come from what we know and from what we can do as researchers, not our personal opinion as individuals' (Bjereld et al., 1999, p. 117, my translation). Methodology helps in distancing oneself from idiosyncratic opinions and personal views and should contribute to more professionally and analytically grounded positions on knowledge claims or research findings. Developing a critical approach of that kind does not happen just like that. As just pointed out, it takes a good deal of expertise to be able to critically engage in a relevant way in a specific field. The same is true of the methodological skills that this book introduces: It is not something you learn overnight. Methodology books of the 'cookbook' kind – also including this one – can give the impression that if you just master a few basic tricks and descriptive models, you've mastered the field. This is not the case. Method is a craft that you learn by practicing it. Just as you don't become a skilled cook by reading or retelling recipes, you don't become a skilled scientific writer or critically thinking academic by reading methodology books. It's really learning by doing. In that process, however, this book and others can be of some help.

***

Methodology is thus about much more than what is (sometimes dutifully) written in the methodology section of theses, dissertations, or scientific articles. It is not just about how to conduct a study in order to obtain high-quality results. It is broader than that. Methodology is something that permeates an argument or a line of reasoning. Knowing methodology also means understanding how conclusions can be supported and being able to engage in a detailed and informed discussion of a knowledge claim. In short, there is an element of general knowledge in mastering methodology.

3

WHAT IS QUALITATIVE METHOD?

Qualitative method is interested in meaning rather than statistically verifiable relationships between variables. This is perhaps the most basic and simple characteristic I can give. At the same time, of course, the answer to the question is more complicated than that. Qualitative method is, of course, also interested in relationships, and just because it is qualitative does not mean that everything that has to do with quantities is irrelevant. If you write that your thesis is based on qualitative method, that doesn't really delimit the thesis very much. So, you must be more nuanced than that. 'Qualitative method' is a term that encompasses many different approaches.

Acknowledging complexity

In his classic *The Interpretation of Cultures* (1973), anthropologist Clifford Geertz underscores that anthropology is primarily concerned with meaning. An anthropologist is interested in how people construct and sustain their worldview and their relationships and thus create meaning. Now, of course, not all qualitative and interpretive research is anthropology. But the idea that what people perceive as meaningful is important to consider, and that reality is seen as constructed by the actors who inhabit it, is a central tenet to many traditions of qualitative research.

Another approach can be found in the sociology of Bruno Latour. In his so-called actor–network theory, it is not so much about meanings as about how different actors influence the course of events and why things develop in a certain way – or cease to develop (stabilize). The point is that different kinds of actors, not only humans but also animals or physical objects, in different ways affect the networks in which they are embedded. It becomes the researcher's task to try to trace what is happening in the network (for an introduction, see Latour, 2005).

These theories (which are quite different from each other) suggest that cause and effect is often complex and sometimes contradictory. The world does not unfold in a simple way before the feet of the researcher. Trying to reduce the complexity of reality to simple relationships is not the task of qualitative methods. Rather, it is to reveal complexity and

DOI: 10.4324/9781003498384-4

nuance. Learning to 'unpack', read, and analyze different social contexts is an important skill to develop through engaging in qualitative research, which will also have implications for life after the thesis.

Quantities

A focus on the qualitative almost always implies a downplaying of the quantitative. But this does not mean that quantitative elements are completely absent in qualitative research.

To begin with, quantity is of course important in terms of whether phenomena are common or not. If a large number of interviewees use the same kind of expressions or recount anecdotes on the same theme, this is of course interesting. If the same behaviors occur in several different contexts that are observed, that is also interesting. It probably indicates that these are not marginal phenomena or that things just happened to be a certain way at a certain time. When we look for patterns in data, this in itself relies on quantities. For the qualitative researcher, however, the precise frequency with which a phenomenon occurs is less important than what it might mean in different social contexts.

Also, the fact that something occurs rarely does not automatically mean that it is meaningless. If you study the life of a religious community, the fact that some rituals only occur once or a few times a year is not a sign that they are marginal phenomena. In many ways, they reflect a number of important things in religious life (for example, they may be seen as symbolic rites of passage or as manifestations of ecclesiastical hierarchy). Or, just because most people only marry once (or a few times) in their life does not mean that it is insignificant. It may even be that the absence of a phenomenon can be significant. If we interview a number of people who work in healthcare about their work and none of them mention patient safety, that in itself may be an indication that this is something worth digging further into.

Thus, quantities play an important role in qualitative research in many cases – but not necessarily the same unambiguous role as when statistical relationships are what is in focus. Rather, quantitative aspects must also be subject to interpretation.

Interpretation

The concept of interpretation is a core concept in qualitative research. At first glance, 'interpreting' may appear to be a rather arbitrary act. You or I can interpret a text, for example a poem, in just about any way we want. And we're both entitled to our own interpretation, right? This seems to be an extremely subjective basis on which to build research.

Now, thankfully, it's not quite that simple. First, there is no single concept of 'interpreting': There is a difference in what interpretation means in for example hermeneutic or postmodern traditions. (For more on this, see Alvesson & Sköldberg, 2018.) Second, and this is key: The point of interpretation in qualitative research is not the interpretation of a single individual trying to make sense of something. Rather, the point is to contribute to a more general understanding of a phenomenon; the contribution is about expanding the interpretive repertoire available. In other words, it is about

FIGURE 3.1 Example of a duck rabbit.

interpretations of a phenomenon that says something of importance to others who are also interested in that phenomenon. Interpretation is thus directly linked to theory and to how research problems are formulated. These, in turn, establish the framework within which the interpretation will become relevant, and it also means that the quality of the interpretation will have a framework to be evaluated against. (These ideas will be explored further in Chapter 6.)

To interpret means to see something as something else. In an oft-referenced example (see Figure 3.1), philosopher Ludwig Wittgenstein (1958) showed how a simple figure can be seen as a duck or as a rabbit, his point being that our immediate perception of the image does not appear to be very much of an interpretation: We see a duck or a rabbit, and that's about it. Furthermore, we do not see the image as just an image or as printed lines on paper. We automatically see it 'as' something, this something at first seemingly obvious and self-evident. But by training our gaze, we can learn to see things beyond the purportedly self-evident, and we can learn to shift between seeing ducks and rabbits. We learn to see the image – or another empirical phenomenon – as something else than what we immediately saw. Which, of course, by extension means that the obvious no longer seems quite so obvious and self-evident.

Interpretative research can be said to be about just this: Developing our way of observing the world around us and contributing to a more nuanced understanding of it.

***

Qualitative method can thus be understood in terms of the ambitions you, the researcher, have for your study, rather than being about using certain methodological techniques.

Although interviews, observations, and ethnographies are common methods (and they will be discussed further in this book), it is not the use of these that is the central point of qualitative research. Rather, what characterizes qualitative research is a certain kind of knowledge interest: In what the researcher wants to say, what kind of research they want to contribute to, and how they want to do it.

4

WHAT IS THEORY?

Traveling on a train, a man suddenly realizes that he is sitting opposite Pablo Picasso – yes, the painter! After mustering his courage, the man leans forward and says: 'Señor Picasso, you are a great artist, but how is it that your art, and modern art, is so strange? Why not paint reality as it is instead of these, eh, distortions?' Picasso then asks the man what he means by 'reality as it is', and the man pulls out a photograph of his wife. 'Here, this is what I mean. Here is my wife'. Picasso looks at the picture and smiles. 'Isn't she very small? And flat?'

Theory – A map of reality?

What is a theory? A common answer is that it is a representation of reality; a common metaphor is that a theory acts as a kind of map of reality. And that sounds good, doesn't it? A map helps us to get an overview, to find relevant information, and to navigate correctly. The map, we then think, mimics reality; it is a representation of reality. This is called the 'correspondence theory of truth'.

But the map is not a straightforward representation of reality. Take a world map, for example. The Earth is (roughly speaking) a sphere, and to make it into a flat two-dimensional map, we have to make some concessions. The world map we most often see is based on the so-called Mercator projection. It dates back to the 16th century and is based on keeping the contours of coastlines constant, which is handy when using maps to navigate by following coastlines. But for the coastline to be correct, something has to give. In this case, the projection leads to a distortion of areas. If you look instead at the so-called Gall-Peters projection (first described in 1855 but popularized in the 1970s), you get a very different picture of what the world looks like – Africa and South America are much bigger and Europe and North America much smaller than the Mercator projection suggests. It's a different world. The two projections say quite different things to someone looking at the map. (See Figures 4.1 and 4.2. If you are interested in the scientific controversy surrounding Gall-Peters and a more nuanced discussion than I have space for here, see Crampton, 1994.)

DOI: 10.4324/9781003498384-5

FIGURE 4.1 The Mercator projection.

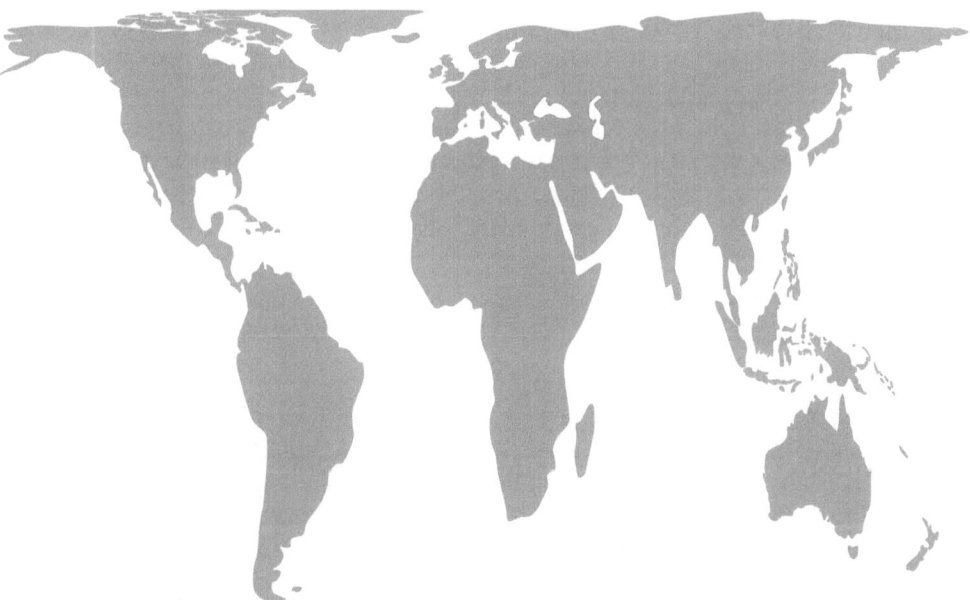

FIGURE 4.2 The Gall-Peters projection.

The map is both small and flat. What many people tend to forget is that a key idea with a map is that it doesn't resemble reality at all. Firstly, a 1:1 scale map would be extremely unwieldy. Secondly, the map does not contain all the information about reality. The scent of flowers and the wetness of water, for example. Of course, a terrain map shows a very different reality from a map of geological or economic conditions or a real estate map. The latter, one might say, is even more real than reality itself, because the boundaries and divisions it establishes exist precisely by virtue of being on the map; they often do not exist in 'reality'.

We always have theories about how things around us work and how they are connected. We may have a theory about why some people drive too fast, about if and how women and men communicate differently, and about why inequalities in society tend to appear. Children may have theories about how to get their parents to buy candy at the supermarket. Theories, then, are not the exclusive domain of science (Corvellec, 2013); neither is empirical data. We live all our lives in the tension between what we want to explain, data, and our explanations, theory (see Chapter 5).

A scientific theory is a complex of ideas. Unlike in everyday contexts, where our theories or rules of thumb are often fairly simple, a scientific theory is a way of 'conceiving the world abstractly, that is in terms of classes of objects and of relations between such classes' (Blumer, 1954, p. 3) and often this results in fairly complex constructions. We can also distinguish between hypotheses (single explanations) and theories (coherent complexes of explanations). However, this is a difference of degree rather than a difference in kind, and in this book, I will use the term theory throughout.

Let me extend the map analogy to theories in general. Firstly, theories do not match reality in every detail. Theories provide an overview and a perspective on things, but not every little part. A theory may provide an overview of how conversations work, but that doesn't mean it will tell you everything about what happened around the Smith family's dinner table last night. Furthermore, different theories show different parts of reality. Gender theory can shed light on what is going on in a conversation just as well as conversation theory, both theories in their respective ways. They will have different things to say about what is going on and perhaps even contradict each other, for example, regarding who has the most influence (partly because the word influence in the two theories is likely to mean different things).

Theory as language

Theories, as Picasso tried to do in his painting, can tell us more about reality than what we immediately see. Theories are a tool of perception, of how we perceive the reality in which we exist. In fact, the very term 'theory' derives from the Greek word *theoria* (θεωρία), meaning 'viewing' or 'beholding'. Rather than understanding theories as representations of reality, theories can be said to offer languages through which we can relate to what we encounter in reality, which shape what we see in reality, and which enable us to share it with others (Brunsson, 1981).

If we see theories as languages, the role of theory shifts. Instead of viewing theories as something that reflects an independent reality, theories become something that creates an understanding of and an approach to reality. Take a concept like 'quality', what does it really mean? Is it an inherent property of a product or service? Or is it what the consumer experiences? By developing theories about quality and what quality means, the very idea of what comprises 'quality' changes. In this way, what we perceive as quality is also changing, and a more nuanced concept of quality provides new possibilities for analyzing and understanding products and services, and it brings new implications for how products and services should be improved.

Or take the example of (social) class: In what sense does 'class' exist beyond the fact that we have a concept to describe it with? There are of course differences between people, for example, in terms of income, education, or identity. But to understand it in terms of class, we need a concept that highlights systematic differences, describes and explores them, and tries to explain why they arise. Class thus becomes a concept that allows us to see and understand society in a different way than if the concept was absent. In this way, it also enables political action.

In this perspective, the outcome of theoretical work is a kind of language development, a way of trying to figure out in which contexts certain theoretical vocabularies are relevant and how they can be made relevant. Sometimes the language may need to be developed, and sometimes the meaning of concepts may need to change somewhat. Sometimes we need to create new concepts and new languages. Research can be said to provide options on different languages that help us deal with the empirical reality we encounter (Brunsson, 1981). By acquiring new languages, that is, new theories and perspectives, you expand your ability to see possibilities and problems in life, and you enrich your repertoire of actions to deal with them.

Returning to the writing of academic theses, this perspective means that it is the role of theory to contribute concepts and models to shed light on what is going on in the empirical material. However, in order to enable theory to accomplish this in a systematic way, it must be related to the specific context of a thesis. It is easy to forget that theory is a central part of the argument of the thesis; it is just not a backdrop. This has implications for how the theoretical content of the thesis is presented (see the section *The role of theory in the thesis*).

Theories and concepts

When it comes to theory, it may also be worth reflecting a little on the use of concepts. Social science is very much built around concepts, that is, words that, in a particular context – for example, in your thesis – have a fairly precise meaning. For example, you might discuss a concept like 'identity' and how it can help us understand prostitution or management consulting. Concepts are often relatively generic and abstract, and thus it is important to be careful about how you use them in a particular context. The same is true of models; they always need to be put into the specific context of a thesis and adapted to it. This is usually done in the theory section of the thesis.

The central role of concepts in social science makes precision in language use very important. This is often solved by providing definitions. Definitions come in different kinds. A stipulative definition means that one decides – stipulates – that a certain concept will have a certain meaning. Once this has been decided, care must be taken to adhere strictly to the stipulated meaning. Of course, it is also good if the stipulated meaning is relatively close to what the term usually means; a stipulative definition is usually about delimiting a term and using it more narrowly than is usual. This gives a precision in the use of the term – a precision which, then, also imposes an obligation, since it becomes important not to deviate from it.

However, this should not lead to an over-obsession with definitions. Concepts in the social sciences are not necessarily easy to define or operationalize precisely. Sociologist Herbert Blumer introduced the idea of 'sensitizing concepts', contrasting this with definitive concepts, the latter referring to more stipulative approaches. Whereas a definitive concept 'refers precisely to what is common to a class of objects, by the aid of a clear definition in terms of attributes or fixed bench marks', a sensitizing concept is not fixed. Instead, sensitizing concepts suggest 'directions along which to look' (Blumer, 1954, p. 7) and will thus change and develop throughout a research project or even a text (e.g. thesis). They are concepts that are allowed to change and shift in meaning. The concept is used to decode the empirical material and it may even be the case that the contribution of a thesis is to show how a concept can be understood and used in a particular context.

The role of theory in the thesis

In a thesis, it is mainly in the theory section that concepts and models are presented in the form of a literature review. But what exactly is it that a literature review accomplishes? One answer, of course, is to review the literature and theories, but this is a rather deadpan answer. However, all too often literature reviews do just that. The author has read up on a number of texts in a field, perhaps by searching databases for what they have seen as key concepts, and then these are described, reference by reference. By this, then you have reviewed the literature. Right?

Assuming that you want to communicate something to your reader with your thesis, what would a literature review of that kind communicate? It communicates, of course, that you have read and understood a number of references. But is that really what the theory section should communicate? Moreover, with the rise of generative large language models, such ways of presenting literature have become extremely easy to do. A language model will do it in seconds. However, this is not very helpful, as it will not assist you in actually understanding the literature, and thus it will not exactly enhance your analytical capacity. Just as critical, it is also a quite pointless way of presenting theory.

The idea of the theoretical part of the thesis is both to explain the state of knowledge about the phenomenon that the thesis is about and to establish one or more central concepts and/or models. The overall ambition is never to simply describe the literature and theories. Rather, the idea is to show how it will be used in the thesis at hand and indicate its role in an overall argument. Although theories are often presented as unambiguous, especially in textbooks, this is not the case. They originate in the tradition of a topic and in scientific debate. To understand theories more deeply, one needs to engage with this debate; otherwise, there is a risk of missing key dimensions, problems, and possibilities of the theory. It then also becomes important to point out to the reader how the debate can be understood in relation to the particular thesis they are reading. The theory is thus a central part of the thesis's argument, as it establishes the 'world' that the thesis is about and how this world can be approached.

What needs to be conveyed is thus not only the breadth of a research field, but also the choices you have made as a researcher within the framework of this field. If we take theories of 'culture', there is a wide range of approaches to what this concept can mean, and it is important to show that you as an author are aware of this. But it is equally important to show what is meant by 'culture' in the present context – it is hardly the case that you in the analysis will work with all the different meanings that the concept of culture can have. (If you try: Good luck.) The theory section is an argument for the relevance of using key theoretical concepts in a particular way in a particular context (i.e. the empirical context of the thesis).

Theoretical background and theoretical framework

The theory section in a thesis can be divided into two different parts that serve different functions in the overall argument.

The first part is a *theoretical background* (or 'literature review'), where the reader is introduced to the previous research in the field of the thesis. However, this review should not cover exactly everything that has been written in the field, but must be focused on what is relevant to the problem the thesis will address. It is all too easy for the text to resemble a textbook or an exam where theories are discussed and explained, but not really related to anything. A thesis, on the other hand, is problem-driven (see Chapter 7). Although a presentation of the existing knowledge in a field should be neutral and factual, it should be shaped to contribute to the discussion of the topic of the thesis and move it forward.

The theoretical background is central, as it describes the state of knowledge to which the thesis contributes. This also means that it is in relation to this theory that the contribution of the thesis is formulated (which in turn is related to the problem background; see Chapter 6). In simple terms, the contribution of the thesis means that you have in one way or another developed the existing knowledge about a particular phenomenon. The theoretical background represents your understanding of this existing knowledge.

The second part of the theory section can be called the *theoretical framework*. This presents the theories that you will draw on in the analysis. This sometimes partly overlaps with the theories in the theory background, but often an additional element is added here that originates in the way the problem was formulated. If your thesis is going to discuss how teachers and students interact in the classroom, previous research on classroom interaction will provide the theoretical background. However, the thesis will probably want to examine a specific dimension of the interaction, such as power or language use. Theories of power (or language use) will then form the backbone of the theoretical framework. This in turn means that the theories of power (or language use) must be related to the previous research on classroom interaction so that the reader can see how it all fits together and how, for example, different concepts relate to each other.

We can see from this that the theory in the paper actually accomplishes two different things for us (here, I partly follow Basbøll, 2023a, b) First, it helps us to describe what we already know about the world – it answers the questions 'What part of the world is this about?' and 'What do we know about it?' That is the theory background. Second, it answers the question 'How can we make sense of this part of the world?' This is the theoretical framework. By extension, this also means that the theoretical background and the theoretical framework can potentially be about quite different theories.

This may sound complicated, but let me illustrate it all from a study I did (Alvehus, 2019a). The study is about leadership; more specifically, it is an observational study about how leadership occurs (or does not occur) in everyday work situations. The theoretical background that the study presents is about the research field of leadership-as-practice. The theoretical background section presents this field of research, but also criticizes it on the basis that it has a slightly too naïve view of power. So here a question arises – how can we bring a more sophisticated understanding of power into the discussion of leadership-as-practice? A partial answer to this comes in the theoretical framework, which introduces Goffman's (1974) theories of how social situations are organized – theories that are not about leadership at all, strictly speaking. But it is Goffman's theory, not leadership theory, that guides the analysis and helps to bring a discussion of power into the leadership-as-practice field. We thus see that the theory of the thesis (the research paper) is about more than simply accounting for existing theories. In this case, it represented a way of describing a state of knowledge, telling about problems inherent in it, and presenting a proposal for resolving this.

In this way, the theory section of the paper becomes something more than 'just' a textbook or exam of the literature. It moves the thesis' reasoning forward and shows how the analysis, and the concepts and models used, relate to the already established knowledge in the field.

***

An overarching point here is that the theoretical part of a thesis is part of the argument that the thesis as a whole represents (see Chapter 6). It is a matter of demonstrating a theoretical awareness, which is signaled by a certain breadth in the theoretical repertoire, and of showing that you have chosen a certain theoretical angle in the thesis. This means that you, the author, must show that this particular theoretical angle is appropriate and relevant in the context and that it can contribute in a good way to the understanding of the problem to be highlighted in this particular thesis. The theoretical part of the paper will therefore be both accounting and argumentative in nature, and the argumentative element is of key significance.

5
WHAT IS DATA?

Empirical data or material is the information that is collected in a research project in order to be analyzed. If, in the context of a study, one collects information about a phenomenon, say, how companies present themselves as socially responsible by reading their websites, these websites comprise the empirical material that one intends to examine on the basis of one's theory. The empirical material one works with should always be directly related to the research question one is asking. There is reason to dwell on this, for it may be more complicated than it initially appears.

First, a brief note on terminology. Traditionally, the terms 'data' and 'data collection' have often been employed. However, that might sound as if 'data' is something objectively given, that can just be collected – as if was mushrooms just waiting to be picked. By the term 'empirical material', authors often want to emphasize that the material is actively constructed by the researcher. The latter is the view of relevance for this book, but I will use the terms 'empirical data' and 'empirical material' interchangeably throughout, mainly for convenience.

What is it that is actually being studied?

Let's say we want to study leadership. We define leadership as 'the influence one person has on the actions of another towards a common goal' – and then we conduct an interview study. Have we then studied leadership according to our definition? Or have we studied how a number of people view leadership? (See Larsson & Alvehus, 2023 for a further development of this argument.) If we are studying corporate policies on responsibility and sustainability: Have we then studied how companies take responsibility – or have we studied how they want to present themselves as responsible in the eyes of the public?

If you study how a company presents itself on the web, what you see on the company's website is objective in the sense that that is how the company presents itself, at least in that medium. This may be different from what actually happens behind the scenes – it may turn out that the company does not live up to its commitments but only maintains a

DOI: 10.4324/9781003498384-6

nice façade – *but that is actually a completely different research question.* Thus, the question

a 'How does [company x] present its social responsibility in public contexts?'
  differs from the question
b 'How does [company x] take social responsibility?'

Both questions are interesting and relevant, but they require different types of empirical material to be discussed. The same applies to the leadership example above. It can be very interesting to study how, for example, coaches in sports view leadership – but this is not the same as studying the process of influence itself. It is important to be detailed about what is actually being studied and how this relates to the question being asked; otherwise, the researcher runs the risk of constructing an empirical material that is not at all related to the question being investigated.

Data and theory

When talking about empirical material or data, one sometimes gets the impression that 'the empirical' is a kind of essence, that there is something that 'is' empirical – as opposed to, say, theory or some kind of text. Take for example a set of interviews or a bunch of questionnaires that have been filled in, collected, and entered in an Excel document. That's empirical data. But what about things that someone has written, for example on a company's website? Or an opinion piece in *The New York Times* – can that be empirical data? Or even a textbook on social work, describing theories and the development of a scientific field of knowledge – it is obviously about theory, but can it also be data?

The concept of 'the empirical', or a definitive distinction between data and theory, is really quite awkward. It implies that these categories are a given when, in fact, it is a relationship. Sometimes the terms explanandum and explanans are used to describe the relationship: Explanandum is that which is to be explained; explanans is what we explain by. (The terms were introduced by Hempel & Oppenheim, 1948 in a discussion of deductive logic.) To take a simple example, where *a* is data (what we want to explain) and *b* is theory (what we explain by):

(1)
a What we want to explain: The child throws a lump of porridge on the floor.
b Explained by: The child is no longer hungry and doesn't want to eat.

A frustrated and tired parent, who knows that the child in question has not eaten since lunch, might see the situation like this:

(2)
a What we want to explain: The child throws a lump of porridge on the floor.
b Explained by: The spoiled kid is grumpy and rowdy and wants to mess with me.

A more advanced variant could be:

(3)
a What we want to explain: The child throws a lump of porridge on the floor.
b Explained by: The child is working on developing an identity as an independent subject.

As the three examples above will illustrate, the explanations can differ and work at different levels of abstraction. But what does this have to do with theory and data? Well, first of all, in most situations in life, we have some form of explanation (theory) that helps us to orientate ourselves in the world. If we compare 1b, 2b, and 3b, we see that we can probably expect different reactions from parents who use one explanation or the other.

However, the key point here is another. Let us assume that, in relation to the above example, we want to make a study of parenting. We might interview a number of parents about how they deal with frustrations in everyday life (such as when their child throws porridge on the floor) and get their views on why this happens and how they act in such situations:

(4)
a What we want to explain: Parents' ideas about causes of children's behavior (1b, 2b, 3b).
b Explained by: Theories of parental identity.

Someone thinks that the child is full and is content with that as an explanation, someone else thinks that the child is deliberately causing problems, and yet another ponders about the child's identity development. What are we studying now, what is 'data'? In 1b, 2b, and 3b above, this is what we explained by, that is, it was three different theories. But now, what was previously explanation has turned into what is to be explained (1b, 2b, and 3b are now part of 4a): Theory has become data!

In the study we want to undertake, we thus need to add a new level, something that can serve as a theory. In the example above, I chose to analyze and explain the parents' statements in terms of theories of identity (4b).

But the example can be extended one step further:

(5)
a What we want to explain: Scientific discourses on parenting over time (4b at different points in time).
b Explained by: The relationship of scientific discourses to societal development.

Let's assume that we find a number of studies of parenting, one from the 1960s, one from the 1980s, and one from the 2010s. We now have three different versions of 4b from three different time periods. We can imagine that, based on these theories, we write a thesis in the history of ideas on the theme '50 years of parenting theory'. In this thesis, then, scientific texts will be the empirical material (5a), and we need a theory, perhaps about how scientific discourses are related to social development (5b), to help us explain it all.

Now, it should perhaps be pointed out that this example is based on a very standardized use of, among other things, concepts of discourse, identity, and the history of ideas. But of course that is not the point. The point is that theory and data are always about a relationship, a tension. There is nothing that unequivocally is 'the empirical', and similarly, there is nothing that unequivocally 'is theory' and that can never be anything else. This is not to mean, however, that all theory or all explanations are equally good or have equal value. On the contrary, it is a central research task – and thus a methodological issue – to be able to distinguish between good and less good explanations and theories.

What we are approaching here is a more relative meaning of the word 'data'. Empirical data or material represents the phenomenon we want to explain – and about which we gather information in one way or another. But exactly what constitutes 'data' depends on the question being asked: What is theory in relationship to one question may be data in relationship to another.

Primary data and secondary data

Sometimes a distinction is made between primary and secondary data. Primary data is empirical material created specifically for the study at hand. Secondary data is empirical material created for another study, but which can be used in the study at hand. If you design a survey yourself to find out the average income in Paris and then distribute it, collect the responses, and record them, this is primary data. If, on the other hand, you get hold of a report from the Paris City Council that lists the results of a similar survey in an appendix, this is, for the purposes of your survey, secondary data.

The difference between primary and secondary data is therefore mainly a question of who collected a data set and for what purpose. In the case above, if you were interested in studying how public surveys present income inequality, the Paris City Council survey itself would be primary data, while the findings in the survey would be rather uninteresting.

What sometimes complicates this issue is how to treat findings from previous studies. Sometimes authors use findings from previous studies as if they were secondary data; it even happens that scientific articles are presented as secondary data simply because they are based on empirical material. But it all depends on how the material is used. If it is the conclusions of the Paris income study that are being used, it is not secondary data; it is just about using someone else's study in building your own theoretical background. For it to be relevant to discuss in terms of secondary data, it must be one's own independent analysis of the data, which usually requires access to the entire data set collected by the other researcher (which is usually not available in report form, especially in qualitative research). Thus, using conclusions from other studies has nothing to do with secondary data but is normally about theory or problem background.

Naturally occurring data

One of the most prominent scholars in the field of qualitative research, David Silverman, argues that qualitative researchers should pay more attention to 'naturally occurring data' rather than 'fabricated data' (Silverman, 2007). By fabricated data, Silverman means empirical material created by the researcher's own activity. For example, interviews are fabricated since they only exist because the researcher orchestrated the interview situation. But there is no reason to automatically assume that what is conveyed during an interview has any unambiguous relationship to what happens outside that situation (see further in Chapter 14). Naturally occurring data, on the other hand, are things that happen regardless of the researcher's involvement, such as meetings (which you can observe) or documentation (which you can analyze as is). The advantage of naturally occurring data is that they become less of a product of the research process.

Another way of saying this is that it is important to pay attention to what kind of empirical data is being collected. If interviews are the empirical basis of the thesis, then the analysis and conclusions will be primarily about interview statements, not about what the interview statements were about. (That is, it will be about, say, 2b and not 2a in the case of the porridge-tossing child.) If the research question of the thesis is about for example parents' thoughts on children's developing subjectivity and agency, their views on these can of course be obtained through interviews, or perhaps even better, via situations where they talk about this without the researcher provoking the answers by asking questions. A study of online discussion forums might shed some light on the issue (see the section on netnography in Chapter 17).

Claims that statements from interviews, focus groups, or similar fabricated situations have direct relevance to contexts outside these situations must be carefully justified. When we conduct interviews, focus groups, or similar, it means that we draw on the help of other people to achieve a first theorization step and then work with the relationship described in example (4) above. In this way, the researcher becomes dependent on the interviewees becoming fellow ethnographers (Mol, 2002). Of course, this need not be wrong – but it is important to remember what one's empirical material actually is about.

***

The data, or empirical material, are a central part of a thesis. It is usually through empirical material that we learn new things, and for many thesis writers, it is the empirical material that is the most exciting part. Finding out new things, how things 'really' are in the 'real world out there', is always interesting, and such empirical curiosity should always be encouraged. Being aware of what empirical data means and what it might represent creates opportunities to write a more initiated and incisive thesis, where the conclusions are clearly anchored in well-composed empirical material.

PART 2

Writing

6

WHAT IS A THESIS?

One way to answer this question is: A thesis is an argument (Rienecker & Jørgensen, 2018). The question that arises from that answer is, of course, what is meant by it. A bit simplified, one can say that the thesis argues

- that a particular problem is interesting;
- that a particular answer to that problem is relevant; and
- that the way in which the answer was reached is reasonable.

In practical terms, this is done by linking the different parts of the thesis in a special way. In order to create a common thread in the thesis, it is not only necessary that the text is coherent from one paragraph to the next, but also that the thesis is coherent on an overall level. This overall structure will be referred to here as the 'U-model', and it illustrates both the different building blocks that make up the thesis and how they are connected.

The U-model

The overall structure of a thesis can be described as a U. This U illustrates the text flow of the thesis, but also the different parts of this text flow (see Figure 6.1). The model is based on a structure for reports proposed by Lekvall and Wahlbin (2001), but I have adapted it for academic theses.

The U-model gives a picture of the different components of the text and what they accomplish in the overall argument. The actual division into chapters may then vary; as a writer, you need to be a bit thoughtful and look at how the thesis can best be presented. There may even be a ready-made structure that you are expected to follow (many university departments have writing manuals that students are expected to adhere to). It may also be worth considering having slightly more descriptive content-oriented headings than, for example, 'Problem Background' and 'Theory' – but again, practice differs between different research traditions and university departments. Whatever the names of the different chapters, the U-model provides a picture of the *basic logic* of the thesis, and

DOI: 10.4324/9781003498384-8

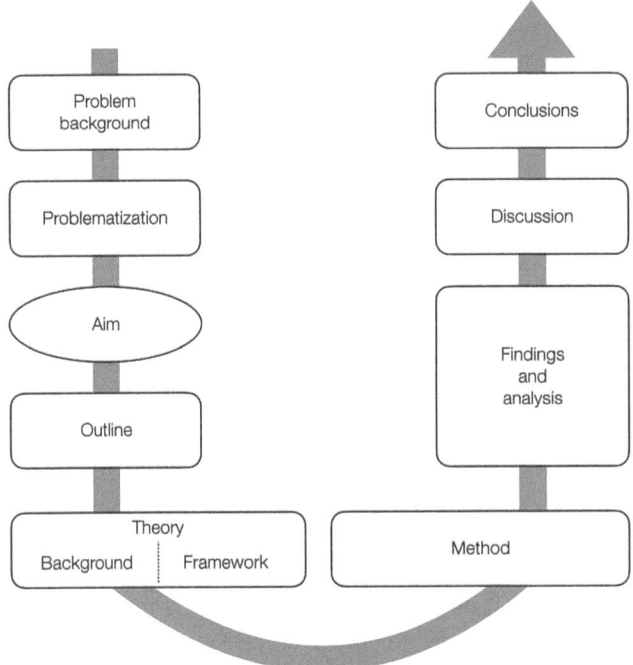

FIGURE 6.1 The U-model and its components.

it is this that is important to understand. If you choose to deviate from the U-model in your thesis, this is fine, but you need to be aware of what you are deviating from, as you are challenging the expectations of the reader.

In the figure, the different components appear to be of the same size, but the picture is not intended to illustrate the extent of individual components or their relative proportions. For example, the 'problem background' and 'problematization' usually cover perhaps 2 to 3 pages in a 40-page thesis. 'Discussion' and 'conclusions' are about the same, perhaps 4 or 5 pages. The main volume of the thesis is normally within what is called 'Findings and analysis' in Figure 6.1. 'Theory' and 'method' share the remaining space, but the theory component tends to be more extensive than the methods component.

Introduction

A thesis begins with an account of the problem background, problematization, purpose, and an outline of the thesis. The problem background describes the phenomenon that the thesis is about and gives a clear indication of the kind of theoretical approach that will be used. In the problematization, the problem is explicated and the precise research question that is to be answered is chiseled out. (For these two steps, see Chapter 7 and in particular the so-called CARS model.) The aim of the thesis is then presented – and here it is important that you are careful! The aim, which is ideally expressed in one sentence, is the single most important sentence of the thesis. This will be examined in detail by the reader.

One can see the aim – or purpose, as it is sometimes called – as a kind of handshake with the reader. Here you make a commitment: *This is what I promise to deliver in this text.* The aim is the primary litmus test for evaluating whether the thesis lives up to its own ambitions. There are then additional requirements for a thesis, but without the aim being fulfilled, the thesis simply has not delivered what it should. (See further on the notion of emic critique in Chapter 19.) It may therefore be good to keep the aim slightly below what you actually aim for and thus give yourself the opportunity to 'over-deliver' a bit. However, this delivery gap must not be too wide, where the aim promises too little. The risk is that the reader will lose interest immediately. Yet, some opportunity to over-deliver is a good idea.

The outline is an underestimated part of the thesis. The outline presents how the thesis will fulfill its aim. It gives you an opportunity to lead the reader's thoughts along the 'right' path from the start, that is, the path that the thesis then follows. In other words, the outline should not only tell you *what* is coming, but first and foremost *why* it is coming and *how* the argument is built up. A well-written outline gives a good picture of how the core argument of the thesis is structured. It is a way of instilling confidence in the reader that it will all work out in the end and giving a sense of what the common thread that they will then follow looks like.

Theory and method

Theory and method are the two fundamental building blocks on which the thesis rests (symbolized in Figure 6.1 by forming the base for the other components).

The theory component presents a discussion of the current state of knowledge in the field, the theoretical background, and the reader is informed of the tools that will be used, the theoretical framework. (See Chapter 4 for a discussion of the difference between these.) Here, parts of the problem background and problematization will reappear, now in a more developed and nuanced form. As emphasized in Chapter 4, it is important that the theory sections do not take on a textbook character, but that they form part of the argument as a whole and lead the argument forward.

The methods component deals with the empirical material of the thesis. Here you explain not only how the empirical material was created (interviews, observations, or whatever) but, just as importantly, why this method is suitable for studying the problem at hand and how the method relates to the theory. The methods section should also include a discussion of analytical methods, which is of course as important as the creation of the material. Both theory and method establish key elements of the paper's overall argument, and ideally, this creates transparency in the line of reasoning. Through well-balanced method and theory components, the two key elements in the analysis – the data and the theoretical concepts or models, respectively – will have been presented and discussed. Of course, for the analysis to be credible, the elements that underpin it must be well worked out – and it is this that will be established in this component of the thesis.

In Figure 6.1, the theory component comes before the methods component in the flow of the thesis. This is not always the case. In traditions where the method has a prominent role, sometimes the methods component comes before the theory component. Sometimes they will partly merge, as can be seen in for example some forms of discourse analysis. As usual, you need to be aware of the conventions of the field you are writing in and relate to them consciously.

Findings and analysis

In the next part of the thesis, the analysis, the basis for the conclusions is created. Sometimes the account for the findings (that is, empirical material presented in the thesis) is separated from the analysis, but more commonly in qualitative research, it is put together into a coherent argument. Otherwise, there may be too many cross-references and it becomes difficult for the reader to follow the connection between the empirical material and the interpretation. The empirical material is sorted and reduced to become part of the argument (see Chapter 18). The idea of giving a complete and 'objective' account of the entire empirical material falls on its own absurdity. Basically, completeness is a practical problem. Apart from the enormous amount of material that would have to be presented, any kind of anonymity would also be difficult to maintain. Moreover, a qualitative approach should not treat the empirical material as objectively given or accessible, as it is from the outset the product of the researcher's interpretations and conceptualization, choice of research problem, and sometimes their interaction with, for example, respondents (see Chapter 20).

In the analysis component, the main argument of the paper is built up. It is here that the reader can follow how you move from empirical data and arrive at the conclusions. This can be done in a number of ways. It will look different depending on the choice of method and on the theoretical tradition within which the thesis is located. The perhaps most important thing is to make sure to stay on topic at all times. Sometimes digressions and sidetracks can be interesting and add value for the reader. But mostly, they are more for the author than for the reader: It gives the author a chance to show their erudition and ingenuity. Such digressions should of course be avoided. There is an expression that is often relevant in this context: Kill your darlings. It is often as painful as it is necessary. The central task of the findings and analysis component of the thesis is to clearly substantiate the conclusions. In other words, the analysis must have a clear direction and move the argument forward.

Discussion and conclusions

The conclusions should, of course, build directly on the analysis. However, this is easier said than done. It is easy to draw conclusions, which may very well be reasonable, without having shown exactly how they were reached. An important step in this is to clearly summarize the conclusions toward the end of the thesis and, above all, to show clearly how they answer the research problem, fulfill the aim, and contribute to knowledge in the field.

In the discussion, you make the connection between the analysis and the conclusions. The outcome of your analysis and your empirical findings are discussed in relation to the problematization that was made at the beginning of the paper. In what way have the empirical findings contributed to answering the research question you have asked? In the discussion, the same references as in the problematization will likely appear again. By entering into a direct dialog with them, you show that you have filled in the gap in knowledge that you set out to remedy, the 'niche' that you have established (see Chapter 7). The discussion component is therefore a reflection of the problematization component.

Similarly, the final part of the thesis, the conclusions, reflects the problem background. The problem background established that the paper addresses a relevant problem. In the conclusions, the idea is to show how your empirical findings have contributed to the broader understanding of this problem. Again, it is therefore beneficial to refer back to the references used earlier and engage in direct dialog with them; in this way, the contribution of the thesis is clearly shown (as it obviously creates a dialog). Often, in reflecting on the conclusions, you can go a little further than where you started, but there is a limit to how far you can go. The broader discussion must correspond to something – and that is what the thesis sets out to unravel, which is already reflected in its opening sections. The thesis should, ideally, end as broadly (or narrowly) as it opened.

Relationships between the components

The U-model shows how the thesis as a whole is connected. Figure 6.1 shows the different components and how they create the overall text flow of the thesis (represented by the U-shaped arrow). The theory and method components are the foundations on which the thesis rests, as illustrated by the fact that they form the bottom of Figure 6.1, but they do not actually provide the main point of the thesis. This is found in the upper parts of the U, where the mirroring between the introductory problem-oriented components and the concluding discussion and conclusion components is central (see Figure 6.2). In fact, we can understand the contribution of a theses constructed according to the U-model solely on the basis of these 'upper' components: Problem background, problematization, aim, discussion, and conclusions. Method, theory, and analysis are there to demonstrate the correctness of the thesis's problem-solving; they do not constitute a contribution in and of themselves.

There are thus a number of relationships between the components of the thesis that you need to take into account. First, all nine components must be internally coherent. Second, they must relate to each other in terms of the text flow (see also Chapter 10). The reader must be led from the theory to the method, from the method to the analysis, and so on. But there are also a number of 'cross-relationships' that need to be established in order for not only the text flow but also the overall logic to be coherent (see Figure 6.2). The

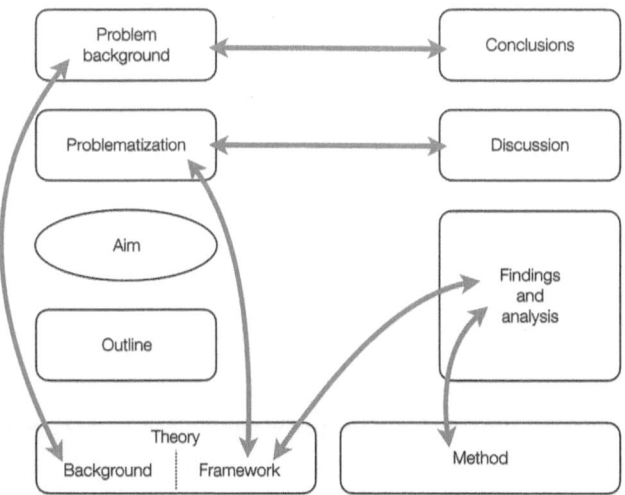

FIGURE 6.2   'Cross-relationships' in the U-model.

discussion responds directly to the problematization: This is where you bring it all together. It is important to ensure that the reader receives a clear answer to the main question and any sub-questions. The conclusions show how the thesis contributes knowledge to the broader phenomenon initially presented to the reader. The conclusions of the thesis are therefore interesting not only for the thesis itself, but may also have relevance for related issues and therefore become useful.

It is also important to ensure that theory, method, and analysis (and, where appropriate, sections containing empirical accounts without analytical elements) are coherent. Figure 6.2 shows that there is a close link between the problem background and the theory background – often the theory background becomes a developed and nuanced version of the problem background. Similarly, there should be a direct link between the problematization and the theories used to analyze the empirical material (the theoretical framework). The analysis should be clearly based on the theoretical concepts and models developed in the theoretical framework of the paper. If you find that there are theoretical framework sections there that are not related to the analysis, these should be looked at extra carefully: Maybe they are not needed? The line of reasoning in the analysis and the presentation of the findings must also match the method chosen. It is easy to write that the paper has an interpretive approach, but then the text must also live up to this and deliver interpretations, not just comparisons between theory and data.

It's difficult to create this full coherence before all the pieces of the puzzle are on the table. This is the source of one of the most frustrating and challenging elements of thesis writing: It is only toward the end of the writing process that everything is visible; it is then that the coherent whole of the thesis can be created. And at that point, time is often in short supply. A good idea is to write as much as possible from the start of the project. It's quicker to delete than to add text (although it can feel more painful). But no matter how much is written, it is only at the end that the whole picture emerges. The only good solution to this problem is to make sure from the start that there is plenty of time for writing and editing at the end. Because no matter how good all the individual components of the thesis are – the methodology, the theory chapter, the account of the empirical findings – they will not come into their own if the whole does not work.

Many considerations

Every thesis comprises a series of considerations. In the writing process and already at the stage when the thesis exists only as a vague idea, perhaps not even written down, decisions about the thesis have to be made. These include

- which question to ask;
- the method to be used; and
- how the argument is to be constructed and substantiated.

What the U-model seeks to explicate is how these considerations interrelate to create a whole – the whole that the reader will face. The aim of the U-model is not, of course, to tie the author to a model that must be followed rigorously step by step, but rather to provide a tool for analyzing one's own writing, both during the process and before final editing, in order to better construct the argument that the thesis represents.

The U-model provides an image of the key considerations that need to be made both before and during the writing process. It also shows how the different components are logically connected (the link between problematization and discussion, for example). It is important that these links work, even if they are not immediately apparent in terms of how the text flows from introduction to conclusion. Using the U-model to analyze other scientific texts, such as articles in scientific journals, also makes it possible to see how different authors both follow the basic logic of the model and make sensible departures from it.

***

The approach I present here may seem extremely convoluted and form-following. That both is, and is not, my intention. On the one hand, theses and scholarly texts are, for better or worse, strongly genre-bound. In order to understand how the genre works, an ideal-type model can be helpful, and that is precisely what the U-model is. As a beginner in a field, it is important to understand the form and how it works. It's no stranger than learning to ride a bike on the street before heading out on bumpy trails. On the other hand, following the form too slavishly risks generating a rather dull text. Yet by understanding and really being able to deal with genre conventions, we can also break them and do this in a conscious and reflexive way. We must also remember that transparency, to which a clear form of argument contributes, is fundamental to scientific texts and should thus be seen in this context as overriding any personal stylistic ideals. Finally, the U-model is in fact similar to Aristotle's (384–322 BC) ideas in *Rhetoric* on how to construct a good speech. The ideal-type form of scientific theses thus reflects long-standing conventions about what constitutes a good and persuasive argument. As I pointed out earlier in this book, method does not only concern scientific texts, but is also a tool that is useful in everyday life. As you learn to understand scientific argumentation, you also learn about arguments and argumentation more generally, an important cornerstone of all critical thinking.

7

WHAT IS A PROBLEM?

The French historian of ideas Michel Foucault begins his famous book *Discipline and punish* (1995) with a detailed and rather gory account of the execution of Damiens, who attempted to assassinate the French king Louis XV. Damiens was carefully tortured, including having pieces of his body torn off with red-hot pincers and having molten lead poured liberally into each of his wounds. The hand that had held the knife was burnt with sulfur. All this while Damien screamed and suffered tremendous agony, but 'he raised his head from time to time and looked at himself boldly' (p. 4). Finally, he was torn to pieces between four horses. Eventually, the executioners had to help carve through flesh and cut tendons to make the body come apart as planned: ' … the executioner Samson and he who had used the pincers each drew out a knife from his pocket and cut the body at the thighs instead of severing the legs at the joints; the four horses gave a tug and carried off the two thighs after them … then the same was done to the arms, the shoulders, the arm pits and the four limbs; the flesh had to be cut almost to the bone … ' (p. 5). The body and its detached parts were burned, and an eyewitness said that 'when they had lifted the trunk to throw it on the stake, he was still alive' (p. 5).

The text is then abruptly interrupted and Foucault goes on to tell us what the rules of a French juvenile prison were like 80 years later: 'At the first drum-roll, the prisoners must rise and dress in silence, as the supervisor opens the cell doors. At the second drum-roll, they must be dressed and make their beds. At the third, they must line up and proceed to the chapel for morning prayer. There is a five-minute interval between each drum roll' (p. 6). The regime is characterized by teaching, work, prayer, discipline, hygiene, and meticulous control of bodies and time.

What is Foucault doing in this passage? From a grim and violent execution where the wrath of society beats down on the body of the condemned and exacts its gruesome revenge – to a discreet, sterile, and total control and regulation of the prisoners' entire existence in time and space. What Foucault wants us to wonder is: How could there be such a complete reversal of thinking about punishment in just 75 years?

This opens up a story about the role of punishment, discipline, and surveillance and how this has changed throughout history – a story that I leave out here. The point as far

DOI: 10.4324/9781003498384-9

as this book is concerned is how in this way Foucault introduces the research problem: He presents something that seems strange and almost paradoxical. How could this happen? As a reader, one is immediately captured: How, indeed? I want to know!

Problematizing can be said to be the core of science. Consequently, problematization is a central part – in some ways the most central part – of a thesis (Alvehus, 2019b).

Welcome to the party!

Writing a thesis is about contributing to the theoretical discussion going on in a particular scientific field, if ever so little. You can think of this as a cocktail party that you enter. At the party, there are already conversations going on between many different researchers. After standing in a corner for a while and looking at the party, it is quite clear that conversations take place in groups, and people are more interested in talking to some than to others. This is where you have to decide which conversation you want to enter to at this particular time. When you have decided, you sneak up to that group and first listen to really understand what the conversation is about. Only then do you introduce yourself and what you want to say, and present what you can contribute. Then, as this is a very nice cocktail party, you get the chance to tell how you arrived at that particular point.

So, before you decide which discussion you want to participate in, you need to get an image of what the theoretical field – the *territory* or if you like, cocktail party – looks like. You don't have to tell us about this in full at once, it can wait until later (more specifically, it is for the theory background component). But in the introduction, you must give a quick overview. You also need to decide fairly quickly who you want to talk to and how you want to contribute to the conversation – what *niche* you want to occupy. Thus, in the introduction of the thesis, the main theoretical references will appear (that is, the names of those who are part of the cocktail party conversation you want to join). Thereby, you signal the theoretical orientation of the paper.

But it's not enough to point out which references you will use – the point is to show that there are good reasons to want to learn more about this particular topic.

The CARS model

If you want to contribute to a discussion, you need to know who you are discussing with and what they have said already – and this is what a *problem formulation* should establish. A well-written problem formulation will ensure that the reader understands what the text is about and what the text wants to contribute. In order for you to show the latter, it also needs to make clear what we already know. One model that describes this is called CARS, 'create a research space' (Swales, 1990).

CARS is based on three steps. The first step is to *establish a territory*. This shows the relevance of the research area. What part of reality will this text address? And why should we be interested in this? The existing research is introduced in concentrate. Step two is to *establish a niche*. Here the reader will find out what is missing in the knowledge of the field. Is there something that needs to be added? Are there theories in the field that contradict each other? Are there reasons to argue that the theory should be understood differently?

As a consequence, comes step three: *Occupy the niche*. When we know what we don't know, we also know what it is that we need to find out in order to improve our understanding of the phenomenon. In this step, an aim is formulated and the thesis also introduces the answer that will eventually be given to the research question. This part of the thesis, the aim and the outline, is central. The outline in particular is usually treated in a too off-hand manner. That, however, is a mistake. The outline should guide the reader through the rest of the text, signaling how everything fits together and showing what answer will be delivered and how.

Sometimes this model is called the 'funnel' or 'hourglass' model (Hill et al., 1982), as the idea is that the problem is gradually narrowed down. It starts in the abstract and general and ends in a concrete and narrow research question – the narrower the better (usually). By this, you also give yourself a chance to deliver a precise answer. However, I generally find the different functions in the CARS model a somewhat more productive approach, as it sets up a tension between problem background and problematization that is very useful.

The CARS model is however rather formulaic and hardly encourages either creative thinking or writing. If you see the CARS model as a method to follow, there is a risk that it will lead to rather conventional research problems. It is therefore important to remember that the CARS model is a form of presentation that need not actually reflect how problems are constructed; I return to this below. However, if you have not mastered the basics of the CARS model, there is a risk that you will not know what in the model you are missing if you deviate. (Foucault's text on discipline and punishment is an excellent example of a well-functioning deviation.) The CARS model is a good way to structure the basic reasoning of a problem formulation. Once this has been done, it is possible to start developing it and making it more stylistically interesting. Perhaps it might be useful to start with an anecdote that illustrates the problem? Or a small excerpt from the empirical findings? A provocation or contradiction that challenges the reader's pre-understanding? Or a personal reflection that the reader can empathize with and that establishes common ground between author and reader?

Four strategies for formulating research problems

There are many different ways to construct research problems. Alvehus (2019b) presents four main strategies for this. It is important to remember what was pointed out above, namely that the way of creating or discovering the problem is not necessarily the same as how the problem is presented in the final text. In terms of presentation, the CARS model tends to dominate and of course this presentation may reflect how the problem was generated – I call this the *gap approach*. But by also acquainting oneself with other ways of constructing problems – the *grounded approach*, the *mystery approach*, and the *actor approach* – it is possible to broaden the repertoire and contribute to more creativity in the process of formulating research problems.

The gap approach is based on the idea that scientific problems are constructed by mapping a research area. The literature is reviewed systematically in order to create an overview of the theoretical field to which one wants to contribute. Once the field has been mapped, it is possible – hopefully – to identify 'white spots on the map', gaps that can be filled in. So, if you work from the gap approach, your thesis process will initially be very theory-heavy, as it all depends on the literature review being done well.

There are objections that can be raised to the gap approach as a general strategy for problem formulation. The first is that the problems created become both conventional and limited in nature. Sandberg and Alvesson (2010) referred to this as 'gap spotting' and argued that research that only aims to plug small holes in theories does not lead to much progress. Another objection is that there may have been a reason for the gap in the first place – it is simply a problem that is not very interesting to investigate further. (On the other hand, it is possible to turn that argument around and argue that it is more a matter of lacking ability to argue for the relevance of that particular problem.) Finally, there is an objection of a more practical nature, which concerns the feasibility of conducting the kind of literature reviews on which the approach is based. There is always an enormous amount of research available and it will be difficult to find the time to map that literature, especially in the context of a thesis process lasting just a few weeks. There is therefore a risk that this approach leads to a rather pointless repetition of what has already been researched.

Perhaps the gap approach could be nuanced somewhat. Often, we think that a gap appears at an empty space and that all other spaces are already filled in once and for all. But if we think instead that it is actually possible to rearrange existing parts of the map and that this new arrangement may lead us to re-evaluate parts of the map that are already there, we get a more dynamic map. (Possibly the map-with-a-gap metaphor has broken down by now.)

> Studies do build on other studies, not in the sense that they take up where the others leave off, but in the sense that, better informed and better conceptualized, they plunge more deeply into the same things. (Geertz, 1973, p. 27)

If we imagine such an approach to finding gaps, there is immediately more space for creative problematization, and the boundaries between the gap approach and other approaches – the three below – become less sharp. Geertz's way of reasoning leaves openings toward less conservative ways of relating to the existing map while retaining the basic idea of adding new research to existing research, albeit in a modified form. New research will not add new supposedly objective empirical findings to old ones, but will use new empirical findings to theoretically deepen, nuance, or even challenge what is already at hand. The newly filled gap thus rearranges the map a bit – but it becomes important and relevant precisely in relation to the map we saw before.

The grounded approach is, in a sense, the reverse of the gap approach. It concerns taking the empirical material as a starting point for formulating the problem. This means that the initial phases of the thesis work are quite different from those in the gap approach: It is a matter of getting out into the field quickly and starting your empirical studies. By getting out into the field and systematically working with empirical material, proponents of the grounded approach argue that problems that are both empirically and theoretically relevant can be created. The approach becomes inherently open to empirical insights and nuances, but at the same time, it brings other difficulties. For example, how can one begin interviewing without having a somewhat clear idea of what to interview about? The interplay between empirical inputs and theoretical insights becomes fundamental, and it raises questions about the balance between theory and data. The approach known as grounded theory (Charmaz, 2014; Glaser & Strauss, 1967) is central to this approach, and within the various versions of grounded theory, there are many ideas about how constructing problems and building analyses worthy of consideration.

A third variant is the *mystery approach*. In the two previous approaches, the problem is based on a lack of knowledge in a field, either based on a theoretical gap or on the fact that something unknown or interesting can be empirically identified. The mystery approach is often (though not always) instead about questioning existing knowledge, overturning or undermining rather than adding to it. Alvesson and Kärreman (2011) provide an extensive discussion of this. The starting point is that neither gap spotting nor data accumulation is a particularly good route to take for someone who wants to say something interesting about society. By using techniques such as defamiliarization (trying to see the familiar in new and unfamiliar ways) or by deeply questioning basic theoretical assumptions, it is possible to work out new theoretical angles that allow us to see phenomena in a new light.

This is where a great deal of theoretical creativity comes into play. If we are interested in planning in a municipality school board, a process that is often seen in terms of more or less rational decisions – why not see this in terms of emotions instead? Or through an aesthetic lens – does a plan get approval due to its beauty? Remember also the potential difference in terms of which theories come into play in the problem background versus the problematization (Chapter 4). This opens up a space for theoretical creativity.

The mystery approach has great potential in qualitative interpretive research precisely because it so clearly works to develop and explore new interpretations. On the other hand, it places great demands on the researcher's interpretive repertoire. While the gap approach is complicated by the amount of literature, the mystery approach offers difficulties because it requires an in-depth theoretical understanding. Truly questioning underlying assumptions, and not just scratching the surface, is difficult. Another problem is that it is difficult to come up with clever and interesting alternatives – a contribution from a mystery approach has to open up new ways of understanding reality in a meaningful way, and that is often quite a lot to ask.

Finally, the *actor approach* takes as its starting point concrete problems experienced by actors. This means that the thesis has the question of relevance resolved from the outset – if the actor can be helped in solving their problem, then the thesis has automatically contributed something. The difficulty here is managing to reformulate the problem in such a way that it also has a broader relevance for other than those immediately concerned. Theses are not consultancy reports. If the thesis is to be of any use to anyone other than the person who originally experienced a problem, then the problem must somehow be generalized and placed in a wider context. The flip side of this is that it can easily lead to the answer being too vague for the actor initially concerned.

There is great potential in this way of reasoning about problems. In recent years, there has been an extensive discussion about the relevance of social science, and the actor approach is in some ways a response to this (Alvesson et al., 2017; Flyvbjerg, 2011; Flyvbjerg et al., 2012).

The four approaches all offer different opportunities and challenges. But the problem formulation is one of the most fundamental parts of a thesis and it is an issue to which you should pay great attention. It is also important to reflect on how the formulation of the research problem relates to both the methodological choices and the theoretical approach of the thesis (Alvehus, 2019b). Just as the previous chapter showed, the thesis must be seen as a whole, where all parts should harmonize with each other.

When the problem is given – By someone else

The image that a thesis always starts with a blank piece of paper and a bunch of theories is thus not true. In many cases, the author of the thesis has a concrete question that, for example, a company wants help with – thus opening up for an actor-type of problem formulation. This must not prevent the researcher from retaining the right to formulate the problem. A thesis should address a problem that is independently chosen by the researcher. A scientific thesis cannot and should not be a purely commissioned job. But there are ways to meet the problem-solving needs of companies or other organizations with the demands of academia to retain the independence in formulating the research problem. This is where the actor approach to problem formulation comes into play, but indeed, there is reason to also keep the CARS model in the back of our minds, as it can help with the tricky business of successfully distancing oneself from the actor's immediate problem.

Say you're in touch with a company that wants help with creating a system to measure their customers' experience of quality. The company wants to get some simple and unambiguous figures to be able to control and monitor the business in this respect. The company may settle for a well-thought-out selection of measurable variables based on research on customer satisfaction. This is however not enough for an academic paper. The question you should then ask yourself is: What kind of problem is this? Is it a question of the effects of measurement and quantification? Is it an issue of quality management? Or is it a question of what customer satisfaction is really about? Depending on what kind of problem you choose to see it as, the territory will be framed a bit differently and the literature you work with will be of a slightly different nature. The next step is to find out what you need to know about quantification/quality management/customer satisfaction. What gaps or deficiencies are there in what we know about this? Are there contradictions in the theories? Based on this, the aim of the paper is formulated and the outline of how the answer is delivered and presented can be made.

By starting from a concrete question, but placing it in a theoretical context and formulating the research problem based on existing knowledge, it is possible to resolve the dual requirements – for example, between a client's wishes and the academy's requirements for theoretical relevance – that often exist. An important point here is also that this can actually help the client to see its problem in a new light.

Yet sometimes this collides. A client may not be happy with you uncovering techniques of manipulation in leadership, for example. Yet again the researcher must own the problem. The relevance of research does not rely on making clients happy.

***

One can say that scientific problems arise in the tension between our pre-understanding of reality and reality as it appears to us when we start to look closer into it (Bjereld et al., 1999). Generally speaking, then, working with a research problem is not so different from everyday life (Alvesson & Kärreman, 2011; Goffman, 1974): We encounter something that we do not quite know how to deal with, and we then try to figure out how to understand the situation. In this way, we change our understanding of how the world works. But research problems are formulated in relation to theory, not to our everyday understanding of the 'nature' of things. Although these sometimes coincide, in a thesis context, it is the theory that is essential. Thus, in order to formulate a research problem, we need to get an idea of what this prior understanding consists of in the research field in question. Research problems are theoretically grounded and have a high degree of precision. A broad and imprecise question inevitably leads to sweeping and imprecise conclusions.

8

WHAT IS A REFERENCE?

Scientific texts can seem impenetrable for a number of reasons. One of these reasons is the references. Why all those names and years that get you stuck in reading, or all these footnotes that make you shift your gaze and lose the rhythm, or endnotes that make you sit and flip back and forth? However, whatever reference system is used, references serve several important functions in a scientific text.

Reasons for referencing

The first important point of referencing is simply to give credit where credit is due. If a person has come up with an idea, produced a piece of research, or contributed valuable insights, that person should of course be given credit for this. Referencing is one way of doing this – and of course, this also means that the author does not take credit for what others have come up with. (Also, researchers are often evaluated not only on what they have published, but also on whether what they have published has been cited. This is however mainly relevant in contexts where publishing in academic journals is the dominant model.) One side effect of this is that references make it possible to trace an idea backward in time. References allow you to demonstrate that you have not just taken the first description of a phenomenon available or the first definition of a concept that you happen to stumble upon, but have actually made the effort to go back and see how that phenomenon or concept developed. Conversely, references indicate that you have taken note of the latest available research on a topic, thus avoiding reinventing the wheel.

Another use of references is to show where facts come from. If you claim that the average income in a particular city is a certain number of dollars per day or that the percentage of middle managers who are stressed at work is 84%, then of course the reader should be able to check this to get an idea of the credibility and relevance of the claim. Perhaps the city has a distribution of income that makes the average income less relevant and the median more interesting? Perhaps the study of middle managers consists of too small a sample or only concerns the private sector? Providing sources for your data increases the transparency of the argument and is necessary for establishing validity in the argumentation.

DOI: 10.4324/9781003498384-10

References can also be used to shorten an argument. Often, authors want to build on conclusions reached by someone else. Briefly summarizing the conclusion and adding a reference gives the reader the opportunity to look more closely at how the conclusion was reached – the method used, the theory behind it, and the structure of the argument. Of course, the author can explain this himself, but in a text based on several different sources, this could lead to a very long and impenetrable text. Referencing becomes a way of shortening the argument while retaining transparency.

Finally, references serve an important signaling function. The references selected show the field of research in which the paper is positioned and the knowledge claims that already exist to which the paper relates (see Chapter 7). The references also signal the type of critique that is of significance. The references with which the paper engages in direct dialog become an important starting point for a critical reading, since it is in relation to these that the paper's arguments primarily must be evaluated (see also Chapter 19).

Ad fontes

But which sources should you turn to? This is not always easy to decide, but as a basic principle, you should have *ad fontes*. This means 'to the sources' and it suggests that, as far as possible, one should seek out the original source of a statement or argument. Sometimes this ambition cannot be maintained. In the next chapter, for example, I quote the Roman rhetorician Quintilian – but the attentive reader will note that I am not referring to Quintilian but to McCloskey (2019). In that case, my knowledge of Latin is extremely limited, and I was not able to track down another English translation of his works in due time, but settling for the one in McCloskey is by no means a problem in this context. Had I wanted to engage more deeply with Quintilian, I would need to track down at least a full translation in order to be able to understand the passage in its context. This was an acceptable exception, but *ad fontes* is the rule.

All too often, thesis writers take their theories from textbooks. The rationale for this is usually that textbooks explain the theory, perhaps more accessibly, and therefore it is just as good to take it from the textbook as from the original source. The problem with this way of thinking, however, is that you are then at the mercy of someone else's interpretation of a source. While it is certainly good to take advantage of other people's interpretations, it is less clever to entirely depend on them. The original source may be inaccurately or insufficiently reported or has been simplified to fit an introduction to the topic. Each time a source is used, it is put into a particular context and thus its meaning will change slightly. The use of secondary references, that is, referring to A only by looking at what B said about A – should therefore be avoided. Of course, how problematic this is varies depending on the type of secondary reference we are talking about; there is a difference between using a secondary reference for a simple factual statement and using it to refer to a somewhat impenetrable and hard-to-interpret Continental philosopher. But either way, it seems unnecessary to rely on a secondary reference – not to mention that the author using the secondary reference will come through as a bit lazy!

Another common type of secondary reference is to get information from, for example, an encyclopedia. These sources of course suffer from the same problem as mentioned above in that they are by definition secondary references. Many also have only anonymous entries, which means that credibility suffers. There is no way of knowing whether what is in the text is the expression of a particular ideological interest, for example. But there are also other, more fundamental, problems. Sometimes researchers take definitions of terms from this type of source with the motive of obtaining a general definition. That's fine if you want a general definition. But often that is not the case. There is a reason why scholarly texts pay great attention to definitions and precision of concepts, as they are careful to be extremely precise in what they say. So, if you are writing a thesis based on a particular concept, you should be aware of the nuances of that concept and exactly how it is used in the particular scientific context. An encyclopedia tries to do exactly the opposite, namely to provide as broad a definition as possible to show the different ways in which a concept can be used. Encyclopedias are excellent for just that, but they are for the very same reason a rather poor source for definitions of concepts in theses.

Citations

How do you write in-text citations? There are plenty of guides on this, both in books and online, so I won't go into detail. The easiest thing to do is probably to look at how a scientific article in the field you are writing in handles citations and simply copy that way. You can also go to the authors' instructions on the websites of academic journals; again, of course, it's a good idea to get information within your own scientific field. Finally, there are sometimes writing instructions or templates available at the department where you are writing your thesis, where you can get practical advice or instructions you need to adhere to.

Broadly speaking, there are two main types of systems that are common in the social sciences. One is called the Harvard system and consists of names and years in the text (as I do in this book). Another is called the Oxford system which uses footnotes. Both citation systems have their respective advantages, and there are several versions of each. The Oxford system can be perceived as more discreet and less intrusive in the text; on the other hand, it automatically means that the reader has to move their gaze to check up on the references. The Harvard system does not have this problem, although it may appear a bit messy with names, brackets, and years sprinkled throughout the text. When using a Harvard system, the footnotes can be used for other things than references which is sometimes useful, if for instance you really can't help yourself from making digressions at times.[1] Different scientific disciplines use different systems. I recommend that you follow the standard for the field you are writing in.

From a writing point of view, I have one main tip: Start working with citations and references from the very beginning of writing. It becomes an unnecessarily tedious task to have in front of you at the end of the writing process when you are often busy getting the thesis as a whole together. You should of course use reference management software, which liberates you from the utterly boring and arduous task of matching citations in the text to the reference list in the final stages of writing.

---

1 Such as this. And just as this one, it is often unnecessary. Apologies.

Also, think about how you incorporate citations in the sentences. Compare these two sentences (both based on the Harvard system):

1 Rigorous methodological standards do not lead to better science (Feyerabend, 1993).
2 Feyerabend (1993) argues that rigorous methodological standards do not lead to better science.

Is there a difference between them? One important difference is that in 2, it is stated that it is an argument; in 1, it appears more as a fact. By choosing one over the other, you shift emphasis somewhat. Another difference is that the citation in 1 is appended to the end and, because it is in brackets, exists outside the sentence structure. In 2, 'Feyerabend' is the subject in the sentence. By writing in this way, it is easier to engage in a direct dialog in the text. This reduces the risk of ending up with mere rehearsals of others' statements and instead encourages a continuous conversation with the literature. This can sometimes help bring a text to life. However, if you only use the second method, you risk ending up with a text in the form 'Lisa thinks this, Karim argues this, but Yasemin says that …', which can be both boring and confusing. There is simply a case to be made for alternating the two methods. Variety is the spice of writing!

Another common question is whether to refer to pages in the citations. There are some guiding rules and principles here. Firstly, if you make a direct quote or borrow a figure, the page should always be cited. The same applies if you take individual details or facts from a text; it may simply be difficult for the reader to find them otherwise. Looking for a single fact in a 400-page book is not the easiest thing to do; so, do the reader a favor and increase transparency by indicating on which page the information can be found. On the other hand, if you are referring to a main point in a text or an argument that is consistent throughout the text, page numbers do not need to be given. That would in fact be quite strange, as the point or argument is built up throughout the text and cannot be found in a single place. The general case is therefore not to give a page reference, unless one of the exceptions above applies.

***

Perhaps the most important point with citations and references is that it helps to create transparency in the text. The reader sees where facts and arguments originate, they see where the text takes shortcuts in its line of reasoning, and how what lies behind the shortcut can be found. In other words, using citations in a clear and systematic way is a key part of what gives a scientific text its credibility. And one day you may even find that you start to miss the citations when they are not there, however unbelievable that may sound to someone confronted for the first time with a seemingly complex referencing system.

9

WHAT IS A SCIENTIFIC LANGUAGE?

Perhaps the most obvious characteristic of scientific language is that it is boring to read. At least, that's an impression you can get if you browse through a few academic journals. (There are exceptions, of course.) At worst, the language is full of long sentences with inscrutably constructed subordinate clauses, long lines of reference, and specialized jargon that, to an outsider, seems both redundant and impermeable. (The former sentence is a borderline case indeed.) But does it really have to be that way? And is it really so?

The importance of clarity

Style is always important in all kinds of texts. As readers, we have stylistic expectations. There is a difference between a chat discussion with a few friends and an opinion piece, a novel, or a scientific article. Different genres simply have different informal rules, and sometimes also formal ones. This does not mean that the rules are impossible to break, nor does it mean that it is not appropriate to break them from time to time. Humor is often based on this kind of 'rule breaking', and in art and music, style breaking is itself an element of style. But at the same time, there are certain rules, or stylistic ideals, that are so fundamental that breaking them is problematic, problematic to the point of risking falling outside the genre. For a writer of experimental novels, a stand-up comedian, or a free-jazz musician, this may not be a problem. But for an academic writer, it might become one.

One such ideal that is worth highlighting in this context is clarity. In Chapter 20, I emphasize transparency as a key requirement for achieving research quality, and this is grounded in clear language. Concepts must be used with care, and lines of reasoning must be comprehensible and possible to follow. Argumentation must be correct and logical. This, in turn, makes issues at a very detailed level (see Chapter 10) important. Sentence structure, choice of words, use of clauses, paragraph structure – all such details must be in place. Like any other text, the research text is about communicating something. The word 'communicate' comes from the Latin *communis*, which roughly means 'common'. Communicating is about the creation of common meanings – we want to make ourselves understood.

DOI: 10.4324/9781003498384-11

Here, however, McCloskey (2019) has an important addition. She quotes one of the great rhetoricians from Antiquity, Quintilian:

> Therefore one ought to take care to write not merely so that the reader can understand but also so that he cannot possibly misunderstand. (Quintilian in McCloskey, 2019, p. 16)

With this in mind, it is important to be extra careful when writing a research text. You can't be satisfied that the reader will probably understand what you mean – you have to ensure that misunderstandings reasonably cannot arise.

It is the reader who decides whether a text is clear or unclear (McCloskey, 2019) – but it is the writer's responsibility (Sword, 2012). An asymmetrical power relationship if ever there was one! In effect, this means that, as an author, you'd do best to seek out readers who can help you reflect on whether your text is clear enough. Critique is a fundamental part of thesis writing (see Chapter 19). It also means that text editing – writing and rewriting the text – is a central and often underestimated part of the writing process. I return to this in the next chapter.

But to return to the introduction of this chapter: Clear does not have to mean boring. On the contrary, I would argue. Texts that are boring are often boring because they are not clear. Of course, if a story is worth telling, if an argument is worth making, it is not made any better by hiding it in complicated sentence structure or in convoluted jargon. Don't worry about making your text 'sound academic'; it's a sickness that runs rampant among researchers and students alike. Instead, think about how you can make your argument as clear as possible so that you cannot be misunderstood.

Can I have an opinion?

Scientific language is characterized by its objectivity and clarity and therefore normative statements and sweeping generalizations are never appropriate in a thesis. (The previous sentence is such a generalization.) This is a common starting point for characterizations of academic writing.

But of course, there must be space for some kind of opinion. In many cases, theses are even written with the intention of making recommendations or suggestions for solutions to a practical problem. And this of course means that an element of opinion, in the sense of evaluative conclusions and perhaps recommendations, comes into play. For example, a study of situations in classrooms might suggest how the psychosocial working environment for teachers could be improved. That a thesis should be unbiased does not prevent such a conclusion.

What is important, however, is that the normative basis for such a conclusion is clearly stated. A conclusion about what is appropriate or desirable cannot be based on a description alone. It is also based on an evaluation against a norm, in the case above, of what constitutes a good psychosocial work environment. By clearly stating this norm – and of course basing the statement on available research – the evaluative conclusion can be drawn. It is therefore not biased just because it is evaluative. Clarity here consists of clearly stating the premises on which the conclusion is drawn and being transparent about these.

What should be avoided, however, is to 'neutralize' evaluative (normative) statements and present them as truths or self-evident facts. For example: 'the overriding objective of a company is to make a profit'. This is often presented as a self-evident fact, but in fact, companies (like other organizations) are characterized by many different goals and stakeholders. For example, management often seems to think that the company's survival is more important than profit; in a health care company, patient safety may be more important than profits (we could hope, at least); trade may consider a good work environment to be as important as return on investments; and so on. If you want to write a thesis based on the assumption that the overriding interest of corporations – to the extent that corporations can have interests – is to make a profit, then of course you can do so, but you must make sure to argue clearly that this assumption is reasonable.

Can I use 'I'?

Another idea about scientific texts, which is related to that of having opinions, is that the language should signal this supposed objectivity. There is a sometimes-expressed idea that one should not write 'I' in an academic text. In some scientific fields, this may be true, but in interpretive research, it has come to be accepted to write in the first person, 'I' or 'we'. Sword (2012) shows – in a limited survey, but nevertheless – that the first-person form is clearly dominant in fields ranging from medicine and evolutionary biology to law, computer science, and literary studies. Put simply, there is often a point in having an author clearly present, a presence that clarifies and reminds the reader of the perspective from which the text departs.

At the same time, there is a danger in the excessive use of the first-person form, and that is that you get a subtle shift in the argumentation of the text. Just because there is an author present does not mean that 'I think that …' is a good argument. If you are arguing that a certain concept should be understood in a certain way in your thesis, then you need to reason about this based on the arguments and references that you can mobilize. (See the example of profit as an overarching corporate interest above.) Sometimes, however, the first-person form causes a slippage, and 'I believe that …' becomes part of the argument. That the author believes in what they write is nice (and one would hope that this is true, in general). But it is not something that should convince a critical reader – the interesting thing, which must be clearly stated in the text, is of course *why* the author believes this or that. It is this 'why' that must not be forgotten; author presence does not make it okay to be sloppy in argumentation!

Metaphors and anecdotes

One way to spice up a text and make it more palatable to the reader is to use metaphors (such as that of food). A well-chosen metaphor, or analogy, for that matter, can be a way to engage the reader and create clarity around abstract concepts. However, it is important to be sensitive about which metaphor. Some metaphors work better than others and there is always a risk that a metaphor you invent yourself will work only for you. At the same time, language is fundamentally metaphorical. For example, a love relationship can be seen as a journey where two individuals' paths cross; they then follow each other through life until the paths fork, although sometimes they may get lost in everyday life and unable to find each other for a while (see Lakoff & Johnson, 1980). Metaphors are a way of making an abstract phenomenon (a relationship) more tangible and understandable (a journey), and this is something you can use.

A variation on this theme is to use examples and anecdotes. If you want to describe complex events, sometimes an example can work in the same way as a metaphor, making it easier for the reader to imagine what happened. The example does not always have to be true; it may well be a fictitious example ('imagine that ...'). The point is not to show how something happens (or has happened) in reality, but to allow the reader to approach the abstract in tangible terms. In game theory, used in economics for example, there is a fundamental problem that illustrates the difficulties of trust and cooperation. It is called the 'prisoners' dilemma' and is exemplified by how two prisoners can choose to rat or not to rat each other out and thus affect their respective sentences. Although the example is completely made up, it illustrates well the delicate balance between trust and self-interest.

Of course, using real-life examples or anecdotes works well too. The example or anecdote must clearly illustrate what is to be illustrated – but not bring in too many other aspects. The point (usually anyway) is to make something clear to the reader, not to stimulate their imagination to go off in all sorts of different directions. As with metaphors, one has to be careful and thoughtful here.

Jargon

Be careful with jargon. Jargon is language that is internal to a particular group (for example, a field of research) and is therefore difficult for outsiders to understand. It involves the use of technical terms and a specialized use of words that have a more general meaning in other contexts. Jargon can thus be used to signify group affiliation. Using the terminology of a field becomes a way of showing that one belongs to that field and that one can communicate with 'insiders' on their terms. However, it also leads to the exclusion of other potential readers.

Of course, there are often good reasons for a scientific field to develop specialized concepts. This is an effect of the demands for conceptual precision that are required of an academic author in order to make a research contribution. But still, be careful. What specialized concepts are *actually* necessary? Are they explained in a clear way that invites the reader to understand? By doing some housekeeping in terms of terminology, the main point you want to convey can be made clear, and perhaps the reader's vocabulary can be expanded or nuanced. But avoid swamping the reader with a flood of jargon that carries the central message away from the reader.

Lapses and blunders

Perhaps we can differentiate between two types of linguistic error: Lapses and blunders (Asplund, 2002). A lapse is a mistake that just 'happens' to occur, a mistake that is certainly undesirable – but it is also no more than that. A blunder, on the other hand, is a mistake that casts doubt on the author's competence and undermines the credibility of the text.

A lapse can be for example a typo or an error in how a sentence's clauses are connected. These are linguistic mistakes that reduce the quality of the reading experience but that do not really affect the reader's understanding of the argument. Asplund gives the example of when he himself unknowingly invented words that did not exist but were still perfectly understandable. On the one hand, lapses are relatively unproblematic, and on the other, they are easy to correct. It is often a good idea to have someone who has not previously read the text do a read-through focusing on language. Together with the spelling function in the word processor and text processing software or plug-ins, most lapses can be corrected. Of course, a single lapse is but a lapse, but too many may become a blunder.

Most spelling mistakes should be eliminated, as they are marked by the word processor. The same applies to elementary grammatical errors, and here the automated systems are getting better and better. However, some mistakes cannot be corrected by spelling software. It gets more complicated when it comes to the specific use of terms that a scientific text may require and when it comes to, for example, misreferences. A reference between two sentences may well be grammatically correct, but it may be completely wrong in relation to what you want to communicate. Computer-based language processing does increasingly help with such issues, but we're not yet at the point where you can trust them entirely. Thus, use the systems, but don't rely on them!

(On one occasion, a student handed in a text to their teacher that was so full of spelling errors – almost every word – that it was illegible. 'How could you miss all that?' the teacher wondered, 'it's all underlined in red on the screen!' The student replied, 'Oh that one, yes, I turned it off. I couldn't read, it was red everywhere!')

A blunder, on the other hand, is worse. A blunder is a linguistic mistake that is not only about language, but also shows that thinking has gone wrong. In the French playwright Molière's *The Imaginary Invalid*, we find a doctor-to-be explaining why opium induces sleep by its 'sleep-inducing power'. This is of course a circular argument that provides nothing in terms of explanation (and Moliére goes on to mock the 'learned' doctors for similar lines of reasoning). Blunders in terms of tautologies, or other similar mistakes, have consequences. A blunder, especially if repeated, is difficult to ignore. The reader loses confidence in the author's argument, and rightly so. The best – perhaps only – way to guard against blunders is simply to get it right from the start. Thinking and reasoning, at least in this context, are language-based, and thus language needs careful treatment.

***

Just because a thesis is factual, it does not mean that it has to be sterile and use dry and neutral language. On the contrary, there is every reason to try to engage the reader in various ways and make your point clear through a well-crafted style. At the same time, of course, this should be done in moderation. Overloading the language with metaphors, using jargon-ridden terms to the point of making the text unreadable, and starting every sentence with 'I …' is obviously not a good strategy. Careful use of language and ensuring that the text is both written and thought out correctly are of course important. It is worth remembering that thesis writing is about writing, and it is about getting a message across to the reader. Making use of the possibilities of language is not wrong.

10

WHAT IS WRITE-WORK?

Let's eat children!
Let's eat, children!
Editing saves lives.

The example is of course stolen from the Internet and perhaps a bit silly and definitely more dramatic than the cases that usually appear in the average thesis. Nevertheless, it points to something important, namely that something as small as a comma can change quite a lot. The devil is in the details.

Too many of the students I tutor are not used to really *working* with their texts. Writing. Reading. Editing. Reading. Repeat. A text is a living entity; it is not just written down and then set aside. In the process of writing a thesis, it is important to start writing from the beginning of the process, even if it is just a rough draft and ideas. Much of what you write will never make it into the final thesis. That's perfectly fine.

I have chosen to make this chapter about 'write-work', not 'writing'. They are, of course, largely the same, but by using the term 'work', I want to try to help you see this as a work process in which the writing itself is but one element of many. I thereby want to encourage a slight shift in focus.

In write-work, there are both smaller and larger issues to be dealt with. As I mentioned earlier in the book, it is only when all the pieces of the thesis begin to fall into place that a whole begins to emerge. This means that a great deal of write-work remains to be done once the thesis is 'finished'. Already when planning your thesis, you need to set aside a considerable amount of time to work on your text at the end of the process. This is not about a 'good to have' time where you do that last polishing of the text. On the contrary, I argue that if we consider the thesis as an argument (see Chapter 6), then the final shaping of the argument is one of the most central parts of the thesis work and should be given the time it deserves.

Write-work must be about both the text as a whole and its details. In the following, I will start with the importance of pre-understanding and the thesis as a whole, and then go into more detail about some hands-on approaches you can apply when reworking your text.

DOI: 10.4324/9781003498384-12

Read fiction!

Sometimes, the idea that academic texts and fiction are completely separate entities is forwarded. According to this view, there is little to be gained from fiction for those who want to learn how to work with specialized academic texts. This is far from true.

Even the most technically oriented scientific text depends on stylistic tricks, on a sense of flow and rhythm, and on an eye for the overall story. At least if it is to be readable and if the text is not to resemble an instruction manual. This could include, for example:

- creating 'hooks' that catch the reader, such as interesting introductions in the form of short empirical extracts, provocative questions, or thought-provoking examples;
- language that is accessible and not filled with obscure expressions, avoiding too much jargon, homemade abbreviations, and neologisms;
- well-functioning textual interconnections, for example, that sentences build on each other and do not appear staccato;
- a clear common thread that both holds the text together and moves it forward.

The text must also be grammatically correct, the choice of words must sound good, and the rhythm must be there. Some of this you can learn from 'how to write' manuals, but a lot of it is about feeling and judgment – things you can easily learn from fiction! Reading fiction is fundamental for anyone who wants to write, also for those who want to write academic texts. Reading fiction teaches us how language works, gives us a sense of the values of words and a sense of rhythm, and teaches us how to construct a good narrative that ties a text together.

To be very hands-on (and perhaps a bit pretentious), here is a tip for those who are reading this book at the beginning of an academic education. One of the best ways to make yourself a good academic writer is to read. A novel (or biography or historical account or ...) is perhaps 350 pages on average and a reasonably skilled reader – as you are or soon will be! – will read about one page per minute without any problem. If you spend half an hour a day reading (instead of doomscrolling), you can read two novels a month. During a three-year education, that's about 70 novels. So, it's quite easy – just make a reading list and get started!

But – and there is always a 'but' – you should of course also be familiar with the academic literature in your field. Find some examples, articles or books, that you have found interesting and worth reading. Think about how those authors have done it; look at how the text is structured. How long are the different sections? How does the argument develop through the text? Where are the main points? How are sentences and paragraphs built up and put together? How are references and quotations handled? And so on. Good role models are important.

The thesis is a narrative

Like any good novel or non-fiction book, a good thesis has a strong overall narrative. There is a clear common thread that connects it all. If you go back to Figure 6.1, the U-model, the overall narrative is about the U-shaped arrow itself. For the scientific logic of the text, the cross-relationships are central (see Figure 6.2). For the narrative logic, for the story, it is the U that is the focus. This is also related to the idea that the thesis is a *coherent* argument.

Perhaps it is in the empirical and analytical part of the thesis that traditional storytelling plays a more apparent prominent role. Depending on how the empirical material is presented (see also Chapter 18), it is often possible to find narrative structures for it. For example, a study that has followed a process over time may simply use time to provide a basic structure. In other studies, it may take a little more work to find a narrative.

Let's say you have identified three different themes in the empirical material. Now you can start thinking about how they are connected. Is it possible to present them in such a way that you go 'from small to large?' Or 'from large to small?' Or is it that in order to see the point of the third theme, we must first grasp the first two themes? Already here, there may be a structure that can be used to create a clear line of reasoning.

Let's say you have studied the everyday work of social workers and your themes are about (1) what they do at work in terms of daily routines and problems, (2) how the management of social services affects the conditions for their work, and (3) how the work is always embedded in collegial processes. One way to create a narrative structure 'from small to large' might then be to start with how everyday life appears through the eyes of a particular individual, how this reflects general patterns, how these are in turn reproduced in collegial interaction, and finally how this interaction between individuals is influenced by the management of the organization. Alternatively, it is possible to start in the broader conditions of work and management, see how these affect relationships between individuals, and end with what this means for the routines and problems of everyday work. One version zooms out, the other zooms in. There will be different narratives, but by choosing one of them, it will be easier for the reader to follow your reasoning because there is an underlying logic that creates the right expectations and constantly hints at answers to the question 'what comes next?'

For those who are well acquainted with qualitative analysis, the question of how themes are related will come up at an early stage. It is familiar and expected. But at the same time, this is a question that goes hand in hand with that final part of the writing process, where all the pieces of the puzzle are on the table. The story you tell in the analysis section must of course be related to your theoretical focus and to the research question, and that will lead you to different conclusions. The overall narrative is therefore not just about the analytical component of the thesis. The theory part, for example, is also part of the narrative and must be framed in the context of the thesis as a whole. The theory background tells you what the state of knowledge is and leads on to a story about what tools might be helpful to fill that knowledge gap or problematize those basic assumptions – whatever the problem formulation was about. The methodology component is the story of how we can find out something about the phenomenon in focus. And so on. If you have found a strong narrative in the empirical material, you might want to keep it on the table throughout the paper. Allow the reader to see glimpses of the empirical material right away – perhaps the thesis can begin with a short vignette that sets the tone from the start? – and return to it from time to time, for example, by reminding of the empirical context. Everything is connected, and since you're writing the text, it's your job to make sure it all makes sense to the reader.

Flow in write-work

I have met people who claim that they write from the beginning of the text to its end in a single sequence. Good for them; I, for one, can't understand how that would work. For me, a text develops in fragments. But the fragments can be of different kinds. For some, it can be very fuzzy:

> We don't write word for word, we don't put one word after another like pieces of Lego. We hardly even write sentence by sentence. The text emerges in fuzzy units that can consist of anything from sharp formulations to vaguely implied, larger chunks of text (Asplund, 2002: 37).

For me, it's not quite that fragmented. I often write one paragraph at a time. I have an idea of what the paragraph should be about and from that I formulate a sentence ('core sentence'). I then expand the core sentence into a 150–200 word paragraph; what I'm basically doing is developing the core sentence to explain and clarify my point. For me, the paragraph is the smallest unit of text I (normally) work with, an idea I learned from the blog *Inframethodology* (a link and some more information can be found in Chapter 22).

The attentive reader will note that this book is not made up of such paragraphs at all. That is true. This is a textbook, not a contribution to a scientific conversation, and I think it works better with a different rhythm in the text. That said, I would like to underscore that textbooks are rarely good models for academic writing of the kind this book is about.

Back to your thesis. Let's say you are going to write a section in the chapter on theory background. You have already read up on theory and taken notes. You already know that it should end up being about one to one and a half pages of text. That's about 1500 words, and if each paragraph is 150–200 words, that means we're talking of about eight individual paragraphs. This means that you can create a common thread for this section from the start by writing eight core sentences that are connected in a logical sequence. Now, the framework is ready. Now it's only – 'only' – about reworking each sentence into a paragraph.

Each such paragraph is based (since the example is about theory) on references (scientific papers, books) that you have read and understood. The paragraph expresses one single idea – the core sentence – in *your* terms, using words *you* choose to convey *your* idea. This is extremely important: Try to never write with the references opened up beside you or on the computer desktop. Do not just rehearse, even by rephrasing in your own words, what the source claims. As I emphasized in Chapter 4, the theory section is not about retelling, but about putting theories into context – the context of your paper's overall narrative. Write what you want to say supported by references, discuss and problematize references – but only recount to the extent that is really necessary for your reader to understand the argument. Avoiding writing with one eye on the source text will force a more independent style and better flow in your argument. This means also that you must know what you are writing about.

For me, this method makes the write-work more manageable: I can break it down into sub-problems, which in turn form the basis of a work process. When I go on to actually write that theory section from my eight core sentences, I do it using the pomodoro method. The method takes its name from the humble egg timer (often tomato-shaped; *pomodoro*). The method is all very simple. You decide to work for a certain amount of time (e.g. 25 minutes) and then take a break (e.g. 5 minutes). The clock starts ticking, you start writing. During those 25 minutes, your smartphone should be in another room or at least in airplane mode, the email should be turned off, all social media should be turned off, and so on. Now you write. Complete focus. For 25 minutes. Nothing can interrupt.

When the 25 minutes are up, the alarm goes off. Now you take a break, stretch your legs, refill your coffee, go to the bathroom, or whatever you want. But after those 5 minutes, you go back into the pomodoro bubble.

The approach is surprisingly effective. I try to write one paragraph per pomodoro session, so I know that that theory section will take 8×30 minutes, which is 4 hours (although I can rarely do more than four pomodoro in a row, then I need a longer break). Voilà! The basic text for the section is there.

It is extremely important to interrupt just as the alarm goes off, even if it is in the middle of a sentence. The point is psychological. When the alarm rings, you interrupt in the middle of writing, at best when you're doing really well. Thus, you stop when you have the pace in your body, you know what you're going to write next – you don't stop because you've got stuck and don't know how to continue. When you start the next session, it's with the feeling that you didn't quite finish, that you have more to say – not that 'oh, here it came to a halt, what do I do now?' It builds a kind of momentum into the writing.

You may have just noticed that I know from the start how long a section or a chapter should be. For me, having an idea of the amount of text helps a lot. It allows me to know from the start how much space a particular argument should take up and it creates a balance between different parts. This makes the final editing much easier.

This approach is quite instrumental and many probably perceive it as very boring. Shouldn't you write out of desire, out of passion? To be curious, to learn and discover things? Sure. To write is to enable systematic reflection. And you're bound to discover things as you write. Plus, you're (hopefully!) writing all the time as you read by taking notes and jotting down ideas. Of course, you should work on your text in a way that suits you – this is my way of working, and others have other ways. My main point here is that it is possible to work in a much more systematic and structured way with writing than many people think and that it's a pretty good way to make sure that you produce text that moves you forward in the process. Because the text isn't finished just because you've written down your eight paragraphs. On the contrary – they are just the raw material from which you will build your text. The basic writing itself is only a small part of the write-work.

Re-reading

At some point in the writing process, you will find there are few things that bore you as much as your own text. It feels like it's been read and written a hundred times at least, and eventually – at least for me – it becomes hard to see what's really there. It's as if my brain all by itself fills in missed words and smooths over grammatical errors. I can be absolutely sure of what I've read, but in fact, it's something completely different that's actually on the screen.

An important element of writing is to now and then create some distance to your text. Critique is a good way to accomplish this (see Chapter 19), and the reader is always the final judge of whether you have been clear or not (McCloskey, 2019). But there is reason to try to work with revising the text both at the end and continuously throughout the process. Let me give you four handy tips for creating a little more distance between you and the text during the writing process, enabling a reading with some fresh eyes.

A first tip is to simply let the text be. By allowing a few days to pass during which you are not working on the text, or instead work on other parts of the text, it will be a little easier to return and read it anew. Unfortunately, in my experience, after a while, it takes a relatively long time for the text to become unfamiliar and it is difficult to get such long breaks in a thesis process. But by working on different parts of the thesis on different days, you can at least create some distance.

Another way is almost a bit too trivial, but I find that it can actually work: Print the text on paper in a different font. Reading for editing on paper is always preferable, I think, but changing the font also makes the text look different and the lines break differently. Some distance is created.

A third idea is to read the text aloud to someone else or to yourself, or ask someone else to read the text aloud to you. This will allow you to almost automatically hear when there is a hiccup in sentence structure or a misreference. In the best of worlds, speech synthesis would be a solution here – it's really not fun to read academic texts aloud! – but unfortunately, flow and rhythm are precisely the major problems with speech synthesis to this end, but that is exactly what we want here.

Finally, quite often theses are written in collaboration. By taking responsibility for different parts of the text (for example, sections or chapters), you can act as readers of each other's text. This automatically creates distance. If you are working from the idea of core sentences, it is quite easy to divide the work – but be careful! It also means that the text will automatically have two slightly different styles, which means that there is still some work to be done to make the text stylistically homogeneous in the end.

Rewriting, revising, and editing

I would say that I spend at most a quarter of my writing time generating text. The rest of the time is spent on rewrites, revision, and editing. It can be a pain, admittedly.

As with write-work in general, I would think that there are as many ways to work with editing, revising, and rewriting as there are competent writers. (For the remainder of this chapter, I'll use the term 'revising' to cover all three activities, just for brevity.) Arguably, one indication of an author's skill is that they are always revising their text. It is of course impossible to know how many endings Ernest Hemingway actually wrote for of *A Farewell to Arms*, but in a posthumous edition of the book, 47 versions are included (Hemingway, 2012). Revising is not only about form, but also content.

Some revision is done continuously in the writing process by going back, rephrasing, or changing one word for another. It is easy to get stuck here, to get stuck on a choice of word or on correcting typos. I usually solve this by skipping that particular word or writing 'xxx' and moving on ('xxx' is easy to search for afterward). I'll find the right word later, and spelling errors I solve in the next step. If you're 'in the text' and it flows, keep going! View the text as a draft, it doesn't matter if something is wrong – it's going to be revised anyway.

But once it comes to that revision, how do you do it? Creating distance from the text is of course important, as mentioned above. But then what?

It is difficult to revise text without a clear focus. How do you revise a 40–50-page thesis? As a whole, you don't. It's all about breaking it down into sub-problems. A good place to start is at the level of the paragraph, as I mentioned above.

Start by reviewing each paragraph and each sentence in it. Read it aloud. Does it flow? Is the punctuation in the right places? Do the sentences follow on from each other in a logical sequence so that the argument keeps moving forward? Are all sentences correct? Is the paragraph really about one single thing, or should it be divided into two separate paragraphs? Are you making digressions that should be removed? Spend at least 10 or 15 minutes on each paragraph; revise if necessary. (It is necessary.) Make sure you are absolutely certain that the point of the paragraph cannot be missed – remember Quintilian in Chapter 9! It is at this level of the text that the foundations of transparency and clarity are laid. If it doesn't work here, it will never work.

Once you have worked through each paragraph in a section or chapter, zoom out a bit. Now it is time to see how the paragraphs fit together. Does each paragraph logically follow the next? Do you need bridges to help the reader understand why you shift to a different theme? Does each section (subchapter) have a clear beginning, middle, and end? That is, does the first paragraph of the section introduce the idea, then the idea is developed in the following paragraphs, and it is brought together and summed up in the concluding paragraph? If not, rework. Add paragraphs if it's necessary, remove if that is necessary. This is where the foundation of the common thread, the overall narrative, is laid. Sometimes you will see tips about working with headings, to have multiple levels of subheadings (2; 2.2; 2.2.1; 2.2.1.3; …), and to have introductions and chapter summaries. In my experience, this does little to actually create the common thread. On the contrary, they often become an excuse not to work on the common thread *in the text*. Look at a novel – do you need explanations of what happens in each chapter and a bunch of subheadings to keep up? Nope. And you don't usually need that in scientific texts either. I'm convinced that good paragraph management means that you almost never need more than one subheading level within a chapter. The common thread must be built into the flow of the text – it cannot be solved with patchwork afterward.

You can do these two steps as the text develops. A paragraph can always be revised on its own, a chapter when it is (sort of) finished. In the third step, it's time for the thesis as a whole. Only now can you see how the different chapters fit together and how the overall narrative actually works. You can look at the cross-relationships (Figure 6.2) and at the flow of the text (Figure 6.1). Now you can start thinking about whether those departures you made from the U-model really work as well as you thought they would. Maybe they force you to make a lot of explanations of how the text is supposed to hang together that don't really add anything? Try cutting up and reassembling the text (but do it in a new version of the document!) to see if other structures work better. Move things around, experiment. Read and re-read.

The fact that you have now zoomed out does not mean that you have left the paragraph level behind you once and for all. As soon as you make changes, they will affect every other part. Any piece of text always exists in a context. If the context of a paragraph changes, the paragraph itself will likely also need to be adjusted, at least a little. The text exists and must be worked with, at all levels simultaneously.

This is why a large part of the time spent on write-work is spent on other things than writing. It's about editing, revising, and rewriting, not generating new words and sentences. It may sound tedious (and it is), but there are some important points to this process.

First, and most obviously, working with the text, and doing it properly, makes for a better end product. The whole of the text depends on the parts, and the parts depend on the whole. The text needs to be worked on from the most detailed level (choice of words, punctuation) to the overall flow of the text and the overall logic of the thesis. A good text will allow you to engage in a conversation making a clear contribution that can be discussed by others and help to move the conversation forward. Remember the cocktail party!

Second, and perhaps less obviously, many people may find it difficult to write because they don't know exactly what they want to say and they fear it won't sound very good. But by approaching the text as a living thing, seeing revising not as a question of failure but as a natural part of how a text comes into being, the writing process is de-dramatized. It doesn't have to be perfect, or even good, from the start. The important thing is that your process produces text that you can then work on, not that that text has a certain quality at the initial stage.

Third, you should ultimately be able to stand up for your text. You are the one who is trying to communicate, to make something common, to convey your thoughts through the text to the reader. Your text should be well thought out. As put by Swedish poet and novelist Bodil Malmsten (2012, p. 130, my translation): 'What remains in the text should be the result of a deliberate choice'.

Practice write-work

There is a tendency to romanticize writing. Perhaps it has to do with some kind of idea of the Great Author and their muse. For the Great Author, the text emerges almost by itself, intuitively, as if it were a matter of 'automatic writing'. Or we see before us the Genius agonizing, tearing their hair, carefully weighing every little comma on golden scales – they have chosen the path of pain.

In fact, writing is very much a craft. Just as we don't get on stage and play improvised jazz without knowing our blues scales or that we ensure we have spent a serious amount of time in the saddle before attempting *L'Etape de Tour de France*, we can't deliver a good thesis without having practiced. As argued by novelist Stephen King (2000), in order to be a writer, there are two things you absolutely need to do: Write a lot and read a lot. There's no way around this. I mentioned reading in the beginning of this chapter – but you also need to practice writing. If you are reading this book at the beginning of your studies: Take the opportunity to start working more systematically on the assignments, essays, and shorter papers you will be writing. You will make progress, I promise!

Also, don't forget that while the dedicated academic author can and should learn from fiction, it is still an academic text that is in focus. In the previous chapter, I emphasized clarity as an important ideal, and in Chapter 20, I discuss transparency as an indispensable criterion of quality. When writing fiction, you can be as liberal as you wish when it comes to such issues. Do your thing. But for a thesis or any other scholarly text, they are not optional. The goal toward which your writing should strive is clarity and transparency for the reader, and working out a good narrative is just a means to that end.

***

Writing is a very personal thing. In writing this chapter, I have drawn heavily on my own experiences as an author. Each author will have to find their own path, which is also part of the point of learning a craft. By all means, have a look at writing guides, especially those that are more specific about language than I have been here, but to become a good writer, you also have to find your way on your own and practice.

PART 3
Fieldwork

11

WHAT IS SAMPLING?

In all types of empirical work, some form of sampling is required. If you are doing a case study, you need to decide which case or cases to study; if you are conducting interviews, you need to choose who to interview; if you are working with focus groups, you need to know who to invite to the discussions; and so on. The underlying principles of sampling may look slightly different depending on the type of method you are working with, and this must always be taken into account. However, there are some sampling strategies that may be useful to know.

Randomized sampling

In questionnaire-based survey research, it is common to work with different types of random samples. This allows statistical calculations to be made in order to generalize. With the help of statistics, we can look at a limited number of cases and allow them to represent a whole population. However, in qualitative research, other types of sampling predominate (although random sampling does occur). There are several reasons for this, not least practical ones.

It also depends on the type of generalization you want to achieve with your survey. Statistically based generalization, which is the point of random sampling, is not the only form of generalization, nor is it the predominant type of generalization in qualitative research (see Chapter 20). This, in turn, makes other types of sampling methods more relevant.

Strategic sampling

The first thing to consider is whether you are getting the information you want from your sample. If you are conducting an interview or focus group study, it is of course interesting to find people who can relate to the issues you want to study. It may also be the case that you want people with specific experiences. This means that the sample has

DOI: 10.4324/9781003498384-14

a strategic element in it; it is specifically designed according to the research questions being asked.

However, there is reason to be cautious here: There is a risk of becoming *too* strategic. For example, many have studied leadership by interviewing managers. But if you want to study leadership, perhaps the subordinates should also be interviewed? And, by the way, is it really true that it is managers who do leadership? Perhaps one should try to find people who are not managers but who are still considered leaders (by themselves or others)? (See Larsson & Alvehus, 2023, for more on specific issues in leadership research.) Any form of strategic sampling will have to deal with these kinds of issues, and it is important not to be lazy or too conventional in the sampling process. Sometimes 'outsiders' can provide as much insight as 'insiders'. For example, if you want to understand working conditions in elderly care, it is probably interesting to interview not only those who work in elderly care but also those who have left the job.

Another type of strategic sampling is when the context to be studied is reasonably well understood already. Let's assume that your thesis is going to be about the working conditions in elderly care and that you are going to undertake a case study of a company that provides elderly care services. In that case, it is possible to quite quickly get a picture of how the company is organized and some key people who should probably be interviewed (perhaps the CEO, the HR manager, and a few middle managers). You may then decide to interview and shadow some of the people who work in the company to find out about their experience of the working conditions and their day-to-day work. In this way, through a well-designed strategic sampling, you can access those parts of the organization that are likely to be interesting to obtain information about.

Snowball sampling

Sometimes you want to study a group that is difficult to identify based on observable criteria (such as through statistics). In a snowball sample, you use those you have already been in contact with, such as interview respondents, to find additional people to interview. In this way, you can search through a network of relevant people and get a shortcut to identifying suitable interviewees. A major advantage of this is that it is relatively efficient (compared to searching randomly). The disadvantage is that you risk getting 'stuck' in a particular network of people who already know each other and thus, probably, have a partially common view of things. Snowball sampling then leads to the research problem not being as broadly explored as it perhaps deserves. And since you've never looked outside the network, you don't know what you're missing.

Convenience sampling

Frequently, different types of convenience sampling occur, that is, sampling based on the type of (for example) focus group participants available. University students are probably heavily over-represented in various types of surveys. Through convenience sampling in experiments, we probably know more about psychology students than we know about any other group of people in history! Sometimes this is not a problem, while at other times it is. Convenience sampling risks reflecting a particular group rather than a broader phenomenon. If, for some reason, one wants to know how medical students relate to various ethical issues, then a convenience sample of a few yearly cohorts of medical students is of course reasonable. But if the sample is limited to students who have taken the same courses at the same university (and who have hung out at the same bars), it risks being too narrow, and it will not be representative of medical students more generally anyway.

It becomes even more problematic when convenience sampling is allowed to provide the basis for the research question. Is it really interesting to study medical students' views on ethics? Wouldn't it be more interesting to study how their views on ethics change as they become involved in clinical work? If the former question, rather than the latter, becomes relevant because of the empirical opportunities (as it is more convenient to sample among students because of access), then this becomes problematic from a scientific point of view (see Chapter 7). The relevance of a question cannot be justified by convenience of sampling.

Numbers and saturation

'How many interviews do we have to do?' is a common question from thesis-writing students. Of course, it is perfectly reasonable to wonder about this when you are about to write a thesis. Whether it is the number of interviews, focus groups, or cases, planning a thesis project is, after all, a highly practical matter. Sometimes the university department also has an answer in the form of a set minimum number of interviews (or focus groups or …) that must be conducted for a thesis at, say, Bachelor's level.

In general, it is difficult to say in advance that a certain number is the right one. For example, it depends on how much information comes out of the interviews. When studying a limited social group, it is sometimes said that 'saturation' can be reached. What is meant is that the same information starts getting repeated and that each new empirical effort (e.g. interview) does not provide much new information. On the other hand, this can just as easily be seen as choosing the wrong interview strategy, where new insights that emerge during the interviews are not allowed to lead to further inquiries and interview questions.

Another question is whether the sample should be homogeneous or heterogeneous. A homogeneous sample – for example, the same 'type' of interviewees or 'similar' cases – makes it easier to make direct comparisons between different people or cases. A heterogeneous sample, on the other hand, provides a broader insight into the phenomenon being studied and contributes to greater nuance. In general, qualitative studies tend to lean towards the latter option, but again, much depends on the research question being asked.

Thus, sampling is an issue that is important for qualitative research. However, it is not important in the same way as for research where one wants to make empirical generalizations for a whole population. In qualitative research, it is not primarily this type of generalization that is relevant, and therefore other sampling strategies than random sampling will become relevant. Achieving 'saturation' or getting interesting insights is arguably more important than conducting a certain number of interviews or observations. This also makes overly rigid recommendations about the number of interviews or similar a bit odd. For example, I and a colleague once had an interview-based paper offhandedly rejected for a conference because the sample was not considered large enough. The paper was based on 102 in-depth interviews ... (I admit I'm still a bit upset about this.)

***

Like with so many other questions, the issue of sampling comes back to the problem that the study seeks to address. The sample must match the problem formulation and research question. This does not mean that the choice of sampling sample is a direct outcome of the way in which the research question is formulated, in the sense that the problem always comes first and that the sampling – or the method in general – simply follows on from it. Sometimes the number of options when it comes to sampling are limited, and then one may have to start at the other end by adapting the research question to the possibilities available to study the issue at all. However, as noted above, care must be taken to ensure that the research question does not become uninteresting because of the sampling strategy.

12

WHAT IS TRIANGULATION AND CRYSTALLIZATION?

Triangulation is a term often used in methodological contexts. The point of triangulation is that by using several methods (e.g. interviews, questionnaires, and document studies), one can fixate one's object of study more precisely and get a better image of it.

It sounds reasonable, but in practice, it is not always that simple. In fact, when using several different methods, you often end up in a situation where the object of study is actually less easy to fixate than it was with just one method. Let's say you want to study shopping among tweens, and you decide to interview some tweens, send out a survey to tween parents, and make observations at a shopping mall. What you end up with as empirical material is: One image of how tweens shop obtained through their own stories, another obtained through the parents' view of it as conveyed through the questionnaires, and a third image that you got by observing them at the mall. The three images will likely point in different directions and provide quite different information. Perhaps both parents and children have different images of what is being shopped and why, and they both contradict what appears to be happening at the mall. The phenomenon of tweens' shopping has not been made any more clear by the methodological strategy, but more ambiguous. How should you handle this material? While it may point in different directions, it is this complexity that the study seeks to capture and therefore it should not be downplayed by an idea of a more fixed object of study.

The triangulation metaphor

The idea of triangulation is based on a method for navigation. In determining the position of for example a ship, three bearings are normally taken, that is, the direction to three points in the environment relative to the ship is determined. Three lines are then drawn to form a triangle, the corners being where lines intersect. This triangle indicates an area in which the ship is likely to be and the size of the triangle therefore becomes an expression of the error in the measurements: The smaller the triangle, the better. The idea of the triangulation metaphor is thus that several methods more effectively capture and more clearly establish the empirical phenomenon. The empirical phenomenon is thus assumed to become more

DOI: 10.4324/9781003498384-15

precisely described and at the same time more completely described by combining different types of material. However, the whole issue becomes more complicated if the triangulation metaphor is examined in more detail. The metaphor is based on an assumption that is not immediately obvious. In order for the three measurements to help identify the exact position of the ship, they must be made in the same coordinate system. In the case of positioning of ships, this is normally not a problem – but in the case of methods, it becomes a problem. It is difficult to argue that a questionnaire, an observation, and an interview deal with exactly the same phenomenon, and thus that the context in which empirical material is produced would be irrelevant and that the different forms of communication employed would be identical. The triangulation metaphor is based on 'same domain' assumptions, meaning that there is an underlying assumption about fixed points in time and space that can serve as stable basis for the triangulation (Richardson, 2000, p. 934).

However, it's rarely that simple. A wide range of perspectives in social science emphasize, in different ways, that what we 'see' in social reality study is at least partly an effect of the theories we rely on. (This can be discussed in a variety of ways and I do not intend to go into detail here but urge those interested to search further on their own; see, for example, the reading suggestions at the end of the book.) Similarly, different methods will produce different kinds of answers to the same question. There is a difference between reading a review of a film and just looking to see if it gets a '4 out of 5' rating. Both types of information may be relevant, but that does not mean they say the same thing.

From a methodological point of view, this means that the idea of triangulation is not at all as useful as the metaphor would suggest. It is not simply a matter of accumulating different types of empirical material, thereby creating a more complete and more unambiguous picture. (As noted in the chapter 'What is data?', it is not certain that different methods even deal with the same phenomenon.) On the contrary, the image that is created by employing several methods is often *more* complex and *less* unambiguous. This is not in itself a problem, but it does mean that the reasons for using multiple methods, and the way in which combining them is treated, need to be reformulated.

An alternative: Crystallization

One way to do this is to change the metaphor. Janesick (2000) and Richardson (2000), among others, propose the concept of crystallization. Just as a crystal breaks up a ray of light into a spectrum, we use different methods to divide a phenomenon into different aspects. However, this does not make the results of the study more certain, reliable, or generalizable. The effect is usually one of increased complexity and it opens up for an increased number of approaches to understanding the empirical material. Multiple methods can contribute to a nuanced and empirically grounded understanding of a phenomenon, rather than increased precision in determining the state of things.

One approach that takes this into account is ethnography, which traditionally uses different methods to approach the field (which can be, for example, a social group or an alien culture). The point of an ethnographic approach is not to determine the content of a culture more precisely, but rather to contribute to a more multifaceted and nuanced image of the culture. The use of several methods for different purposes comes in handy here (see Chapter 17).

Let's get back to our shopping tweens. You might observe them talking about shopping in a break between lessons and manage to record this. After that, you interview the three participants. Most likely, this will produce three slightly different images of the conversation, and it is impossible to prioritize any single one of them. In addition, you will have a recording of what was actually said. But this is not the same as saying that what was perceived by the participants in the conversation is exactly what was said (the difference can actually be quite significant). The different methods used here offer the possibility of producing a more nuanced and multifaceted picture of the interaction. What was 'actually said' will differ depending on what material we look at, as the meaning a participant makes from a conversation will depend on a lot more than the actual words spoken.

***

Triangulation can be a misleading metaphor, and perhaps the most problematic thing is not that an empirical material based on several different methods becomes a bit scattered; the really problematic thing is when the analysis is based on an expectation that all the parts of the material say the same thing. The use of multiple methods should be characterized by nuance and multifacetedness rather than by an expectation of a more 'fixed' object of study – and therefore crystallization is the better metaphor.

13

WHAT IS A CASE STUDY?

Case studies are common in many different types of research. What constitutes a case can vary in scope. In economics or political science, countries or regions may be cases; organizational researchers often consider an organization or a department as a case; in psychology, an individual patient may be a case. It is therefore difficult to give a general and simple definition of what actually constitutes a case. Stake (2000) describes a case as 'one among others', that is, an empirical entity of a type of which we can identify several and then select one: One country among several, one organization among several, and one patient among several. The case is a distinct system with its own identity, a distinctive pattern of behavior or system of meaning that distinguishes it from its surroundings. The individual case also (often) points to larger contexts and the idea of describing a case is often to illustrate a more general phenomenon. This does not, of course, preclude the use of 'extreme cases' to study just that – extremes – or to explore where the boundaries of 'the ordinary' lie (Pettigrew, 1990). A common critique that can (and sometimes should) be leveled at case studies is the extent to which it is possible to make empirical generalizations from them.

Thus, while there is an element of generality in the case study – 'one among others' – the case study is also a study of the particular and of the idiosyncrasies in the specific case. An important point of studying a particular case in detail is to be able to look at the specific conditions of that case. A tension between uniqueness and generality is thus inherent in the very idea of the 'case'.

Cases and sub-cases

Already in the premise of the case study, there is an assumption of some kind of whole. A case is defined from the outset by drawing a line between what is the 'case' and what is the 'environment'. Furthermore, a case can often be divided into sub-cases. If the focus is on the organization of a municipality, the case study potentially includes a wide range of sub-cases, such as the city council, the HR support staff, or the parks and recreation department. If one wants to study the municipality as a whole, one will have to choose a relatively high level of

DOI: 10.4324/9781003498384-16

abstraction; if one wants to make more detailed descriptions, for example, of how individual working groups are affected by political decisions, then the case must be further specified. It will then be appropriate to focus on a sub-case. A number of decisions then need to be made: What to observe, who to interview, and so on. The choice of these, in turn, depends on the case as a whole. If we want to understand the organizational conditions in the parks and recreation department, we will have to take the specifics of its connections to the city council into consideration: Who talks to whom, et cetera. The overall assumption of what comprises the case is central to determining clearly what is being studied and therefore it also determines the scope of what becomes possible to say.

One advantage of a case study is that it leads to a concentration of the empirical material. For example, the organization of work in the parks and recreation department can be studied in its specific context, and different statements about work or relationships relate to the same events and people. In this way, for example, conflicts will become concrete empirical phenomena instead of being attributed to a more generic and abstract level. The case provides an opportunity to create realistic descriptions of, for example, organizational processes.

A case of what?

Anthropologist Clifford Geertz once wrote: 'Anthropologists don't study villages [...]; they study *in* villages' (Geertz, 1973, p. 24). An analogous argument can be made with regard to case studies. That is, one does not study a case, one studies *in* a case. An interpretive study, based on a theoretically grounded problematization, will of course not study exactly all aspects of a case. First, this would of course be quite impossible. But second, and more importantly, it is not in line with a well-specified aim of a thesis. This means that the case must be studied from a particular angle, with a particular research problem in mind. You would study a certain phenomenon in a particular case in order to learn something about it; it is the phenomenon, not the case as a whole (whatever that may be), that is of interest.

This does not mean that the point of case studies mentioned above – that they enable a holistic approach and an understanding of a phenomenon in its empirical context – is lost. On the contrary. Approaching the phenomenon *in* the case precisely means that it can be put into its empirical context and it can be related to other phenomena.

It may therefore be worth asking the following question when studying a case: A case of what? Even if you decide to study the collaboration between two organizations, 'customer service' in health care, or how marketing rhetoric works in sales meetings, the question remains: A case of what? In the first case, is it about inter-organizational relations? Or about trust-building between individuals? Or about how different types of organizational relationships become formalized? All options are there, but they lead to different questions and to different theoretical approaches. A case never speaks for itself: It has to be expressed in theoretical terms in order to be interesting in the thesis. A fundamental problem here, however, is that this can be difficult to know in advance – one of the ideas of a case study is to explore what the case is all really about, which in a way makes the case study a particularly complicated challenge. It may also be worth pointing out that the case study is an approach that often appears in connection with actor-oriented forms of problem formulation (Flyvbjerg et al., 2012; see also Chapter 7).

Like other types of empirical studies, the relevance of the empirical material thus depends on both the problematization and the theorization. Already in the first step of the CARS model (see Chapter 7), it is important to keep the case in mind. After all, it is in relation to this 'territory' that the case should appear as relevant to study. In reality, when writing theses, the case often comes before the problematization. Case-based theses are often written on the basis of questions and problems that organizations have, and contact with the case is established at an early stage in the thesis process. However, even if the case has been the starting point for the study, the case must be made relevant to the overall argument of the thesis, and this must be accomplished in the introductory parts.

Multiple cases

Sometimes it can be interesting to conduct multiple case studies. One reason given for this is that it would improve the potential for generalization (Eisenhardt, 1989; Yin, 1994). The problem, however, is that in order to reach levels that are statistically interesting, a very large number of case studies must be carried out. The possibility of making empirical generalizations about the entire populations of 'nursing teams in health care' is unlikely to increase (in a statistical sense) if one goes from studying 1 team to studying 4 or 12, even if some anomalies and idiosyncrasies can perhaps be eliminated.

However, there are other reasons for conducting multiple case studies. For example, if we compare two nursing teams, one working in the public sector and the other in the private sector, the different organizational environments that the teams work within may give opportunities for interesting comparisons. Another approach might be to compare a team where employees are very satisfied with their work situation with one where they are dissatisfied. If we combine these two dimensions, we might consider studying four cases (satisfied + public; satisfied + private; dissatisfied + public; dissatisfied + private) and thus create even more opportunities for comparison. It is of course possible to imagine an almost infinite number of set-ups for this type of comparative case study. In one study I worked on, after going through a few variables, we were quickly up to over 30 cases. We were studying schools and we could, for example, take into account location (big city, medium city, small town), the school's performance in terms of student grades (e.g. dividing it into high or low performers), the level of education (e.g. elementary school, middle school, high school), and whether or not a school was located in an area with a high degree of social challenges or not. Multiplying these options gives us $3 \times 2 \times 3 \times 2 = 36$ possible cases. In the end, we chose to conduct 7 case studies using our variables to achieve some breadth in selection; it was impossible to conduct over 30 case studies with any kind of relevant depth.

The point with multiple cases is generally not to increase the number of cases in order to increase statistical generalizability, but rather to expand the space of possible interpretations. That must, of course, again be related to the research question one wants to investigate. Multiple cases sometimes provide opportunities to ask different questions than single cases do.

However, these possibilities come at a price. If there are limited resources available (for example, only two weeks of observations in total), then multiple cases lead to each case being studied more superficially. As depth is one of the main points of the case study, loss of depth must always be set against the potential benefits of multiple cases. Unfortunately, there is no general rule to rely on in this, but you, being the one conducting the study, must make the trade-off and argue for it. In the school study just mentioned (see Alvehus et al., 2019a, b, 2021), we were three researchers who worked for over two years to create the empirical material – far more time than most student theses allow for.

***

Case studies are common and for a good reason. For qualitative researchers, they are a convenient way to get closer to a phenomenon and study it in depth. Sometimes, however, I get the impression that it is an all too easy choice to go for an interview-based case study. In many instances, a broader interview survey might yield equally interesting results, or why not a netnography? That said, case studies have an important role to play in qualitative research, but it is important to carefully consider the choice of a case study approach as well as the choice of case to be studied, and to ensure that they both respond well to the problem formulation.

14

WHAT IS AN INTERVIEW?

Interviews are, possibly together with ethnography, one of the flagships of the qualitative method. Interviews appear to be an almost indispensable method for finding out how people think, feel, and act in different situations. Therefore, qualitative research is very often based on interviews and frequently exclusively so. In many ways, the interview offers a powerful tool for the qualitative researcher, since by interacting with their respondents, they can ask about feelings and motives, find out how different people view a sequence of events, or how a particular phenomenon is presented in stories and anecdotes. However, this does not mean that interviews are unproblematic as sources of information.

Different kinds of interviews

The first question we should ask is, of course, what an interview is. For example, does it have to be a direct face-to-face interaction between interviewer and interviewee? In the most common case, one might think of the interview as something that takes place in direct face-to-face interaction, but both audio-only and video interviews are of course possible and provide opportunities for interviewing respondents that would have been impossible to meet 'in real life'. The argument can be extended. Can you interview someone by email? Or through a chat? I once heard a professor claim that he had interviewed over 500 people in a project, which of course sounded impressive until the professor had to admit that he had 'interviewed via questionnaires' …

As usual, it is difficult to provide comprehensive definitions. In the following, I will primarily discuss face-to-face interviews, although most of what I discuss can be applied to telephone or video interviews, too. But for example email and chat interviews fall outside the scope of this discussion. However, as I just indicated, this is not a given demarcation, and several of the aspects discussed as follows also concern these latter forms of interaction.

Interviews are not a single method. How interviews are used may differ depending on whether one is analyzing specific events, people's life stories, or people's recollections of service experiences on the recent Thailand holiday. It is therefore important to describe

DOI: 10.4324/9781003498384-17

the type of interviews conducted in some detail: Are they for example in-depth interviews of 2–3 hours or are they shorter, more thematic, interviews?

Miner or traveler?

Why conduct interviews? One good reason is that interviews are a way of finding out how another person thinks and feels about a particular topic, event, or phenomenon. Interviewing a person is a way of getting at their opinions, feelings, experiences, and thoughts. There is much to this, and in many ways, interviews play an important role in creating an understanding of how individuals and groups construct and reproduce their social world.

Qualitative research has long romanticized interviews as a way of getting 'closer' to respondents and generating more credible answers than, for example, surveys. Murdock (1997, p. 188) argued that 'with all interpretive studies, there is a strong residue of Romanticism, a desire to dig down, below surface conventions and elisions, to reach a solid bedrock of experience'. Of course, one should not romanticize interviews too much – I will address some problems as follows – but the fact remains that interviews are one of the primary and most common qualitative methods for accessing people's subjective experiences and opinions.

It is however important to see the interview as a specific form of social interaction that takes place in a specific context. Sometimes you get the impression that interviews are about going 'out there' and collecting data – whatever golden grains of information you can get your hands on. You might even get some inspiring or controversial quotes (real golden nuggets). Against such a mining metaphor, Kvale (1996) contrasts a travel metaphor, which is about the interviewer embarking on an educational journey where they allow their world to be reshaped through interactions with others, as the explorer and the explored together create an insightful experience. This metaphor also has strong elements of romanticism, but offers a more interactive view of how interviews work, underlining that they are about co-constructed images of reality, not about ready-made data that can be collected. However, in different ways, the two metaphors are about taking part in other people's life worlds and reaching some form of understanding of this.

There are researchers who take a more pessimistic view of the interview's potential to represent something 'outside' the interview situation (see for example Alvesson, 2003a). Interviews are then seen as local phenomena where what is said at the time of the interview has little or no significance outside that particular situation; or that the narrative created during an interview is based on (unconscious) schemas and models of what should be said in an interview and how it should be presented; or that the interview is an arena for identity construction, where the interviewer and interviewee present fitting (and flattering) images of themselves to each other. This objection should be taken seriously, as it reminds us to be careful about what it is that representations actually do represent.

Structured, semi-structured, or unstructured?

In terms of conducting interviews, an initial problem faced by the interviewer is how to make the conversation happen. How many questions to ask, for example? How much space should the respondent be given to develop their answer? Who – interviewer or interviewee – should set the agenda? For example, when should you interrupt to get the conversation back on track? And who should decide which track is the right one?

In the mining metaphor, the interviewer already knows what they want. They set the agenda and ask the questions; the interviewee answers dutifully. The traveling metaphor is more open. The traveler (interviewer) has a goal with the interview, but the way of getting there is not obvious and even the goal is negotiable. It thereby becomes relevant to talk about the degree of structuring of the interview. Interviews can be fully structured, with predetermined questions and sometimes also predetermined response options. The interview will then be similar to a questionnaire. This typically makes the interview shorter and more superficial (a 'talking questionnaire'). The advantage is that several interviewers can conduct the 'same' interview with many different respondents. However, an overly structured interview takes away much of the point of interviews, namely the interactive and following-along elements.

In a semi-structured interview, which is perhaps the most common, the interviewer has an interview form of a few open-ended questions or broader themes around which the conversation is centered. Here, the respondent has much more opportunity to influence the content of the interview, and the interviewer has to be more active in listening and in working with follow-up questions. A completely unstructured interview is essentially an open-ended conversation guided by an overarching subject interest, where the interviewer is brought to the background, possibly mainly contributing encouraging phrases and gestures. This does not mean that the interview ceases to be an interview; it is still to be regarded as a rather special form of interpersonal communication and the interviewer will still, through subtle gestures and responses, influence the interaction.

Most interviews fall somewhere in the middle of this spectrum and can be considered semi-structured. The interview still requires an informed and skilled interviewer to come to its right. In order to be a productive conversation, the interview must not degenerate into an interrogation or some kind of test of the respondent's knowledge. Being interviewed should not be about being held responsible or being tested on one's knowledge of a topic. On the other hand, the interviewer must not uncritically and passively absorb all that the respondent says. Sometimes more incisive questions or provocations on the part of the interviewer are appropriate. It is also a good idea to be somewhat restrictive about the number of questions you bring to the interview; otherwise, there is a risk that the interview will be more about making sure you get all your questions answered than getting to hear the interviewee's story. Often a few themes, prepared by reading up on the theory and being aware of what you want the interview to be about, will suffice.

Asking and listening

In an interview, the interviewer asks questions. It may sound simple, but it's worth giving a little extra thought to what you ask and how you ask it. In the semi-structured interview, it is not only the questions on the interview form that are important. It may not even be necessary to have a prepared set of questions. Often it is more a matter of asking follow-up questions and encouraging the respondent to keep talking. This may involve asking the respondent to elaborate on their answer ('Can you tell me more about that situation?') or going into more depth ('How did you feel then?'). Sometimes it is useful to ask about generic and impersonal topics to get a conversation going ('Can you tell me a bit about the culture of the company?') but often it is beneficial if the questions relate to the respondent's own experienced reality. (Kvale, 1996, has examples of kinds of questions that are good to master.) In general, it is useful to learn how to get the respondent to talk as much as possible. After all, it is their story that you are after, not your own.

An important skill for an interviewer is therefore to actively listen. The interviewer must be prepared to follow up on interesting leads, perhaps asking the interviewee to go deeper or to ask for examples. Moreover, a good listener *shows* by posture and gazes that they are interested in what the interviewee has to say and in this way encourages the interviewee to talk further. Silence can be an important part of the interview. By daring to be silent when the interviewee has finished a phrase, you give them a chance to think and perhaps tell you more. If you can withstand the awkwardness of silence, this can be a great way to encourage the interviewee to continue talking.

The interviewer's questions (and silence) will influence the direction of the interview and influence what is said. As mentioned above, it is important to pay attention to the interaction process of the interview. This is true not only during the interview but also during the interpretation and analysis. The emergence of certain themes over the course of several interviews may just as easily be due to them being 'provoked' by an interviewer, who has an interest in the area, as to their importance to the respondents (Rapley, 2004). Most often, when interview excerpts appear in accounts of empirical material, only the interview responses appear, but it may be worth including the questions as well to show why a particular response appears (see, for example, Czarniawska's, 2011, study of news agencies for an example of this).

Transcribing

Once the interviews are conducted, they usually need to be transcribed. Either they have been recorded, in which case there is a daunting task of transcribing them, or notes have been taken and should be filled in as soon as possible afterward.

The advantages and disadvantages of recording interviews can be discussed. Recording may disturb the interviewee and be perceived as limiting their openness. In such cases, it may be necessary to rely on note-taking. The risk of taking notes is the inevitable process of translation: That which goes onto the notepad is not always what was said, but what the interviewer heard. It is also difficult to write fast enough to get all the necessary information and at the same time be a good listener. Sometimes it can be reassuring for the interviewee to be recorded, as they know that what is said will be correct, word for word. Whether the interview is audio (or video) recorded or recorded by note-taking, it is the interviewee who must have the final say on how the interview is recorded.

Transcribing is the first step in the analysis. Here, speech is transformed into text, and this in itself is a step of interpretation (Ochs, 1979). There are a variety of ways to transcribe, and the variant you choose depends on the type of analysis you want to be able to do. In conversation analysis, everything is transcribed very carefully, from intonation to line breaks and pauses. For other types of analysis, it may even be appropriate to tidy up the language a little and make it more like written language. Sometimes AI-supported software can be of great assistance, but you still need to listen to the interview while reading the transcription to ensure accuracy and to make notes of nuances such as tone of voice (for example signaling hesitation or irony). Ultimately, the way in which you transcribe depends on the type of analysis you will make, and there are more detailed discussions of this in the more specific methodological literature.

***

Interviews are an effective and powerful tool in the qualitative toolbox. However, remember that interviews are a very special kind of conversation. This calls for some caution in the use of interviews and other types of 'fabricated data' (Silverman, 2007) – a critique that is just as relevant for focus groups or some types of observations. Although interviews can often feel authentic and sometimes become almost therapeutic, this does not necessarily mean that they represent the way in which things are expressed in other contexts. On the other hand, it does not necessarily mean that what is said in an interview is completely unrelated to what happens outside the interview. Interviews can also be a way of allowing voices that are not expressed in other contexts to emerge (Rapley, 2004). Interviews are an important method in qualitative research and in many ways create opportunities in that they provide access to other people's ways of understanding the world. As always, when it comes to methods, there are strengths and weaknesses, and the methods literature encourages us to be attentive to and aware of these.

15

WHAT IS A FOCUS GROUP?

Focus groups have become increasingly common as a research method. They have been around since the 1940s and have become a popular tool not only for researchers, but are also used in other fields, such as marketing. (The first are said to have been analyses of audience response to US war propaganda; see Halkier, 2008.) Although focus groups are in many ways similar to interviews – situations arranged to find out people's subjective views and opinions on something – there are reasons to treat them separately. A key characteristic of focus groups is that the process is as important as content; *how* things are said is as interesting as *what* is said. In the following, I will focus on the former aspect, as the latter is so similar to interviews, and what was written in the previous chapter largely applies.

Why focus groups?

A focus group is a group of about 10 – recommendations vary, but usually lie between 6 and 12 – people who come together for a limited time (often 1–3 hours) to discuss a particular topic or issue under supervision. The group is led by a moderator, who ensures that the interaction flows as desired and that the group sticks to the topic(s) that are to be discussed. What is particularly interesting about focus groups is the group interaction. By using focus groups, we can analyze not only the opinions, positions, and attitudes expressed during the focus group, but also the processes that are involved in shaping meaning. Thus, focus groups provide insight not only into what attitudes exist, but perhaps more importantly, how attitudes are formed and developed during social interaction (Krueger & Casey, 2008). The role of the individual becomes less prominent in a focus group and there is a risk that individual, differing views are not heard, and it is of course not possible to know what has not been said.

Attending to the process aspect of the focus group means that what was presented as problematic in interviews (see Chapter 14) in some ways becomes an asset in a focus group. It is the interaction between people that is the point, which means that the special situation is something that is useful rather than problematic. Gibbs (1997), for example, argues that focus groups have a higher degree of authenticity than interviews, because

DOI: 10.4324/9781003498384-18

they are less controlled by the person conducting them. However, it should be remembered that the focus group is as odd a situation as the interview is; it is not a 'naturally occurring' conversation that takes place. Again, it is appropriate to be wary of what kind of generalization (see Chapter 20) one makes from the findings in a focus group study. At the same time, the focus group provides an opportunity to study processes we are otherwise not always aware of. Most of us do not give much thought to how our opinions and attitudes are shaped by the social contexts in which they are expressed. The focus group provides opportunities to gain insight into that process, and the process itself is not necessarily that different between the workplace lunchroom and the focus group. Processes that we are not even aware of are, of course, more difficult to consciously manipulate and manage (which certainly does not mean that they are not unconsciously adapted between contexts).

Conducting the focus group

In order to conduct a focus group, you must first decide what kind of questions and topics you want the group to discuss. These must in turn be adapted to the research question of the study and be operationalized so that the focus group can engage with them. Asking questions about 'pastoral power' (Foucault, 1982) in leader–follower relationships will likely not yield very much insight into what you want to unveil; you would be better off asking the participants to talk about the role of care and of showing consideration. The questions must be introduced and presented to the group in such a way that the participants can understand and relate to them. It is often suggested that it is useful to start with more general questions and gradually become more specific. In this way, the participants' attention is not directed too much by the moderator in the early stages, and the participants' associative capacity is utilized better.

Another thing that needs to be managed is the selection of participants for the focus group. We need to consider for instance heterogeneity versus homogeneity and the existing social relationships between the participants. To start with the homogeneity/heterogeneity issue. Let's say you want to investigate how gender is constructed when people talk about fashion advertising. You want responses from women and from men, and perhaps from different age categories. The question is, should you have one group of younger women, one of younger men, one of older women, and one of older men – or should you put together four mixed-gender groups with representation from all different age categories? Depending on your choice, the interaction in the groups will probably be different, and the findings will be slightly different.

The participants' social relationships will also influence the interaction in the group. A group of people who are strangers to each other will act differently from a group where the participants already know each other (and thus have a common past and future). If you are using focus groups to study the culture of a company, for example, the interaction will most likely be affected by whether you have participants from different hierarchical levels in the same group. It may be difficult for subordinates to speak openly in front of their bosses, but on the other hand, the focus group provides an opportunity to study such effects of hierarchy and how for example status is accomplished. That said, the topic (such as 'company culture') is not always that relevant per se, but acts as a trigger for studying processes that are of interest (e.g. status and hierarchy).

As in so many other cases, there are no general guidelines here, but these decisions must be made in relation to the study objectives and also in relation to the possibilities of attracting participants to the focus groups.

The role of the moderator

The moderator has an important role in the focus group. This is both because they must be familiar with the subject that is to be discussed and because the moderator must understand and be able to manage the group process. This does not mean that the moderator necessarily needs to have a certificate in focus group moderation, but some training and preparation is certainly an advantage. Situations may arise where the moderator needs to decide who gets the opportunity to speak or when some individuals need to be held back to give others space. Strong emotions and even aggression may be involved, and it is important that the moderator is mentally prepared for this and has strategies for dealing with situations that may arise. An important overarching decision also relates to when the moderator should intervene to get the group back on track; how and when to decide what is a relevant sidetrack and what is not?

In this way, the moderator will have a strong influence on the outcome of the focus group, both in terms of content and process. This is an influence that, like all other ways of influencing the construction of empirical material, must be handled consciously both when it comes to analytical work and generalization.

Recording the interaction

The activities of the focus group – what is said, who says it, how the participants react to what is said, and so on – must be documented in one way or another. One way to do this, of course, is to record the conversation and then transcribe it (see Chapter 14 for a few words on transcription). It is worth remembering that transcribing this type of situation, with many different voices sometimes speaking into each other's mouths, is in itself a time-consuming process. Again, automated transcription is of help, but it is certainly not to be entirely trusted. In addition, if video has been used for recording, there is a large amount of non-verbal information to work with. This is of course not a problem in itself, but it needs to be taken into account at the planning stage of the thesis, as such transcription and analysis will take quite a lot of time. Another way to record the focus group is to work with a group memory. For example, the moderator or a co-moderator can continuously write down what the group comes up with and expresses on a flipchart or whiteboard (which can be photographed and then erased when full). Again, this will affect the process in some ways, for example, the group may become more consensus-seeking than is perhaps desirable (wanting to 'agree' on what to put on the whiteboard). It also focuses more on the outcome than the process. A further variation is to have an assistant keep continuous notes of what is going on. An advantage of this over only audio recording (or as an addition to recording) is that body language and other reactions can be registered. Relying entirely on what the moderator remembers is of course also possible, but this makes the information very uncertain and thus the analysis shakier; it is generally not recommended.

Process and content

Analysis of focus group data is, roughly speaking, about two things. First, it is about the content of the focus group discussion. In relation to this, one can work analytically in the same way as one works with interviews or observations (and as usual with the method's inherent possibilities and limitations in mind). Secondly, and this is a bit of the point of focus groups, the group process itself can be analyzed. Here one is more interested in how and why different attitudes or meanings are expressed. In order to work with this type of analysis, the researcher needs to acquire knowledge of, for example, group dynamics, conversation analysis, or some other appropriate framework. Through this, the analysis can approach, for example, how identities are expressed in conversations about gender and fashion and how different kinds of positioning between identities are created and reproduced in the interaction. This is of course something that also needs to be connected to the problem formulation of the thesis and the theories that are elaborated in the theoretical framework.

***

The focus group method, like interviews, is a way of accessing opinions and attitudes. While interview studies often focus on the content of what is said, focus group studies are often also interested in the process. (Here, interview-based approaches can probably learn a thing or two from focus group approaches, since interviews also have a process dimension.) This also means that focus groups and interviews are two methods that are relatively easy to combine, not least because they produce similar empirical material. Focus groups can of course also be combined with other methods (but see Chapter 12).

16

WHAT IS AN OBSERVATION?

Observations have a long tradition in the social sciences. Many classic works in anthropology and sociology are based on long-term participation and observation of cultures and social contexts (see also Chapter 17). There are also examples of more structured forms of observation, such as experiments. Observing human behavior and interaction is thus an important part of social science, and in many ways, observations are a way of escaping the problems of 'fabricated data' that plague many other methods (Silverman, 2007).

'Naturally occurring' situations

The reason for doing observations is often to study 'naturally occurring' situations. This can be tricky in many ways, and one of the more fundamental problems usually cited is the so-called observer effect: That the observer influences what goes on in one way or another, making the observation less representative of what naturally occurs. That we are affected by the presence of others has been known for a long time and was in some ways the origin of a long tradition of studies of social psychology. Of course, this cannot be ignored if one wants to study 'natural' situations. An observer will in one way or another influence what happens in a situation. For example, we can probably assume that the presence of a researcher or a recording device will affect what is said during a performance appraisal or similar feedback situation.

In interviews and focus groups, the presence of the researcher has very distinct effects, as the researcher is there to shape the situation. In relation to this, the observer effect in observations appears to be relatively limited. After all, a passive observer does not initiate and try to control the situation (as in the case of an interviewer or moderator). The performance appraisal would probably have taken place even if the observer had not been there, and much of the content would probably have been rather similar. Interviews and focus groups are by definition fabricated situations, and this must of course be taken into account in the analysis. The point of observations, on the other hand, is that they are observations of naturally occurring events. And then the observer effect becomes more problematic.

DOI: 10.4324/9781003498384-19

There is a risk that this 'observer effect' is exaggerated. This prevents researchers from doing observations and instead relying on for example interviews – as if they were more natural (which they are of course not). In reality, the observer often tends to be forgotten by participants, at least for some time, as many social situations are quite involving. Having made extensive use of observations in my own research, I have many times had situations where participants have afterward noted that they forgot I was there. Of course, that doesn't automatically mean I was not affecting the situation at all, but at least it doesn't seem to have been too impactful on the participants' attention.

Moreover, this depends on what kind of observations we are talking about. There are for example observations with different degrees of participation, from the completely passive observer, the 'fly on the wall', to the fully participatory observer, who tries to become part of a group's activity. The extent of participation is one of the dimensions that must be considered in observational studies.

Overt and covert observation

Another dimension is whether there is an overt or covert observation. In an overt observation, the observer makes themself known to those being observed; in a covert observation, they do not. This choice raises various problems, both technical and ethical in nature.

The ethical problem varies between different types of situations (Lofland et al., 2006). To what extent does a researcher have the right to take part in other people's lives and activities without their knowledge and consent? A study based on covert observation should be preceded by careful ethical considerations. Covert observation in private settings can be seen as jeopardizing individual trust and as an exposure of the private sphere. Observations of public contexts, such as studying how people interact in a café, are likely to be less problematic. It is also possible to make observations in the form of self-ethnographies in a setting (e.g. a workplace) to which one normally belongs (Alvesson, 2003b). In such a situation, one needs to reflect on what is appropriate to disclose and what is not. Although covert observations raise ethical concerns, their value should not be underestimated. The German journalist Günter Wallraff has on several occasions studied social environments incognito, thus revealing conditions (e.g. discrimination) that would otherwise not have been made visible. 'Wallraffing' in this way allows the observer to go behind the stage of the theater of public life. Ethical demands on research to be based on informed consent do indeed make this quite impossible.

There are also technical problems with the covert observation that should not be underestimated. For example, it can be difficult to take notes; the field notes become highly dependent on the observer's memory. Depending on what is to be studied, this can be more or less problematic.

Open observation is easier to manage ethically and technically, but as noted above, the observer effect must then be taken into consideration. Now, an observer effect is of course present also in covert observation, because the observer interacts with those being studied. However, this is more of a 'participant effect' rather than an 'observer effect'. In some cases, the observer effect may be stronger, in others weaker. As mentioned, in my personal experience, people seem to forget quite quickly that there is an observer present. Again, we need to take the specifics of the study into consideration. Language use and interaction patterns may be more difficult to adapt to the researcher's presence than specific types of information (e.g. gossip or badmouthing someone).

Choosing situations and contexts

As in any other empirical effort, observation studies require that a selection of observations be made, taking into account what is accessible and appropriate to study in order to answer the research question. Often observations can be the outcome of simply being in a particular setting for a period of time, 'hanging out and asking questions', as Dingwall (1997) puts it. Observations can then take on a distinctly informal character.

Observations can be of a more or less ethnographic type. Ethnographic observations (see Chapter 17) are based on the observer trying to become part of the social environment being studied over a longer period of time. Other observational studies involve more specific types of situations (for example, doctor/patient interactions or project meetings). In these cases, observations often need to be combined with other methods in order to put them into context. An exception here is some forms of conversation analysis, in which the context is only relevant to the extent that it is referred to during the interaction (see Garfinkel, 1967).

Shadowing

Another form of observation is shadowing (Czarniawska, 2007). It involves following a person in their work or everyday life for a limited period of time: A day, a few days, a week … A key element is that this to some extent enables the researcher to challenge the constraints of time and space that otherwise provide limits to observations. Shadowing will also involve interacting with the shadowed person and discussing interpretations of situations, for example, before and after a meeting. Feelings and experiences can be discussed in direct relation to specific situations. The flip side is of course that it is mainly one person's point of view – that of the 'shadowee' – that is presented (although shadowing can also leave room for, for example, short interviews with other people).

Shadowing has the advantage that you can access situations that the people you are studying, and perhaps yourself, would not otherwise study. On one occasion, I shadowed a manager for the purpose of studying leadership. At one point during the second day of shadowing, he was about to leave his office, and when I got up to follow him, he said it wasn't necessary, as he was just going to have a brief chat with some of his subordinates. I chose to follow anyway, and during the small walk-round he took among the staff, he accomplished more leadership than at any other time during the shadowing and observations I conducted. He himself apparently did not consider this to be leadership, as he did not think I needed to follow and he knew it was leadership I was studying. (See Sveningsson et al., 2012, for an interpretation of this.)

Experiments

Experiments are a quite unique form of observation. In social psychology, different kinds of experimental studies have provided a lot of exciting information about how people act in different situations. For example, Stanley Milgram's famous experiments on obedience to authority provided insights into the darker sides of human nature. In organization theory, the Hawthorne experiments provided important insights into how organizations work and the importance of group and social factors, although this was not actually what was originally intended to be studied (Roethlisberger & Dickson, 1939/1947). In studies of justice and in economics, experiments are used to study how individuals prioritize in decision-making.

However, there is always a danger in generalizing from experiments. Sometimes, the experiment is so cleverly rigged that the experimental design itself is actually exactly the kind of situation one wants to make a statement about. Milgram's (1975/2005) experiments are an excellent example (but see Korn, 1997, for an interesting account of the use of deception in social psychology). In other studies, the experiments only remotely resemble a 'real' situation. For example, experiments in different areas of expertise have shown that it is not always so easy to determine what actually constitutes expertise in a particular field. In tax law, experiments have shown that more experienced specialists do not necessarily give better advice from a legal point of view than less experienced ones, but that the two groups differ in terms of how aggressive they are, that is, how far into the gray areas of legislation they are willing to go in their advice in order to minimize taxation (see e.g. O'Donnell et al., 2005). The problem is that in reality, the client is the one who primarily determines the degree of aggressiveness, and it is the competence in dealing with clients that is crucial for success rather than purely legal competence (Alvehus, 2017). Thus, the experiments are only marginally relevant when it comes to understanding how tax advice is produced.

***

Observations are not any kind of 'unmediated' or 'pure' form of empirical material. The very act of observing involves selection and interpretation: Where is your gaze directed? (If you look at the person speaking in a meeting, you don't quite see the others.) Who is being shadowed? Which cafés are visited? Furthermore, of course, not everything will be recorded, partly because not everything can be written down and partly because the observer will already have some theoretical preconceptions and will thus observe some things more easily than others. What types of narratives influence that which is written in the field notes, what metaphors are chosen, and what impressions are selected? Language, as always, shapes research, for better or worse. Yet observations are extremely important in the social sciences and deserve their place as a key method for qualitative researchers.

17

WHAT IS ETHNOGRAPHY?

It is not an easy task to describe briefly what ethnographic method is all about. Generally speaking, ethnography is a way of approaching different social groups and cultures in order to try to create an understanding of a group's or culture's world view and way of life, using as open and broad a methodological repertoire as possible. In many ways, ethnography represents a kind of ideal of qualitative method. It is based on the researcher participating in the life of a group and in a sense becoming part of it, the researcher uses interviews, observations, informal conversations, documents, artifacts – indeed, anything that can facilitate the understanding of the selected group. Early ethnography was encyclopedic in nature and was primarily concerned with 'alien cultures'. Considerable effort was put into collecting, documenting, and classifying objects (in an almost linnaeusian manner). Over time, however, anthropology became more about trying to understand other cultures, what Malinowski (1922) called 'the native's point of view'. A central starting point for ethnographic approaches is to focus on the attitudes and meanings of 'the Other'. Ethnography is thus closely interwoven with anthropological approaches.

Over time, the ethnographic method has migrated from anthropology to other parts of the social sciences and has been used in various ways and to a different extent. In sociology, for example, the so-called Chicago School has been influential, and in the field of business administration, the interest in corporate cultures in the 1980s gave ethnographic methods a boost. Ethnology studies a wide range of cultural expressions, from backpacking to board rooms. The 'alien cultures' to which the researcher travels may therefore as well be found around the corner of the next block, as on a sun-drenched island in the South Pacific.

Writing ethnography

Ethnography can also be seen as a way of writing research and I doubt that any other methodology has problematized the role of writing to the same extent. What is the role of the researcher/writer in the research process? How should this presence and author-ity be handled in the text? Van Maanen (1988) argued that ethnographic accounts can be of several

DOI: 10.4324/9781003498384-20

types involving differing author's positions. They can be realist (downplaying the author's presence), confessional (where the author tries to come to terms with and problematize his or her own subjectivity), and impressionistic (where the author's subjectivity is taken more or less for granted and the aim is rather to use it to produce interesting stories).

Another common ambition in ethnography is to give voice to groups that would not otherwise be heard and to bring subjugated forms of knowledge to the surface. Often such studies take class, gender, sexuality, or ethnicity perspectives as their starting point and seek to make visible the practices and patterns of meaning that influence societal processes. Herein lies another stylistic challenge, that of being able to talk about 'the Other' in a way that is revealing without at the same time reproducing conventional categories (Fine, 1994). How, for example, does one write about rape victims without at the same time risk reproducing their identification as victims – and thus undermining attempts to regain power and the right to define one's own social identity?

The stylistic presentation and treatment of the Other in ethnography will of course have a major impact on how the knowledge and insights produced will be interpreted. (And this applies not only to ethnographies, of course, but also to qualitative research more broadly; see Chapter 9.) It is not surprising that ethnographies, with their ambition to approach the Other closely and with great sensibility, have had to grapple with these problems thoroughly.

The 'big' ethnographic project

Ethnographic work often involves long field studies in direct interaction with those being studied. Six months and more is not uncommon. Ethnographic projects are therefore difficult to plan and predict in detail. This is of course true of qualitative research in general, especially if it is of a more exploratory nature, but in ethnography, it becomes a guiding principle. As Van Maanen (1979) puts it, it is as much about drift as it is about the design of the research project. That means that an ethnographer must be prepared to follow the course of events and follow up on clues that emerge in the course of the research; not everything can be planned and predicted.

Observations, interviews, and conversations play a central role as the ethnographic researcher seeks to gain insight into and understand how a group of people interact and what their customs and worldviews are like. In ethnographic fieldwork, the researcher's initial question and interest only partially set the course. It is important to get a feel for the empirical context and be prepared to change focus and follow along in the stream of events. In terms of problem formulation, a grounded approach, or a version of it, often becomes relevant (see Chapter 7). It is thus only natural that an ethnography is characterized by some drift in relation to the initial research question.

Often, however, this type of major ethnographic project, involving months of fieldwork, is more than can be expected in an undergraduate thesis, as the time for such a project is normally more limited. In what follows, therefore, two less extensive variants of ethnography will be introduced.

Microethnography

An alternative to a full-blown ethnography is to conduct a microethnography (Alvehus & Crevani, 2022). This involves selecting a smaller group, or perhaps even a single individual, to follow over a shorter period of time (cf. Pink & Morgan, 2013). As in ethnography, several different methods – interviews, observations, and shadowing – are used to create as broad an understanding as possible. Since the possibility of following along in the drift is reduced, it is also an advantage to have a slightly more specific focus from the outset than ethnography normally has. Microethnography also does not necessarily have culture and meaning as its primary interest, but is primarily concerned with generating 'thick descriptions' (Geertz, 1973) that are relatively limited in time and space. The primary interest becomes not so much trying to capture cultural wholes, but rather a more concentrated way of decoding, for example, local cultural expressions or capturing practices and patterns of action as they happen.

The relatively tight demarcation in time and space means that microethnography has certain representational advantages, as the empirical material is easier to present – there is simply less that needs to be reduced and taken away. Taken to its logical extreme, microethnography can even be built on vignettes which themselves constitute the empirical material. Interpretation and analysis are then about opening up interesting and creative interpretations utilizing theory. Since what is interpreted is only the material presented, a more symmetrical relationship between the researcher and the reader is created.

Netnography

Another alternative to the traditional 'physical' ethnography is to study communities online – to do a netnography (Kozinets, 2010). If you want to study the ethics of trad climbing, you could of course track down some climbers and interview them. And you will probably get rich material about ethics (it is quite a big issue in trad climbing). But perhaps the question can be illuminated in an even more interesting way by seeing how such discussions take place in 'naturally occurring' situations, when different arguments are actually voiced? We could then imagine that you observe climbers when they interact, perhaps at the National Climber's Association annual meeting or at a mountaineering course in the Alps. But there is more readily available empirical material if you go online and study discussions on social media and other forums where climbers meet. Interaction on social media is, of course, a genre in itself – but so are annual meetings and chats by the fire in a cozy Swiss Alp hut. And there's nothing to say that one is more authentic than the other.

Unlike situations where the researcher creates the empirical material themself (see Chapter 5), the web often offers material that is authentic in the sense that it was not created specifically for the purpose of the study in question. For example, if one studies discussions on an open Facebook group or in a public discussion forum, then these are discussions that are 'naturally occurring': They have taken place entirely without the researcher provoking them. Moreover, the so-called observer effect (see Chapter 16) can be completely avoided, as the person collecting the material may simply lurk: Just read without interacting.

Yet, netnography does not need to be hidden. Quite on the contrary, it can often be useful to make oneself known to and interact with the community being studied in the same way that a 'normal' ethnography is based on a high degree of participation. Kozinets (2010, p. 108) argued that participation is in fact key to netnography; removing this removes an element of embeddedness that is key to ethnographic endeavors. When participation is minimized, so is the opportunity to develop an understanding of cultural practices and meanings together with those in the community.

A clear advantage of netnography is therefore that access to the material is immediate and also that much of the material is already in text form, which makes it easier to analyze (and no transcriptions are needed). A further advantage is that the material is public, which means that the material as a whole is open to scrutiny. This is of course also true of the more participatory parts of a netnography, if they take place in public forums. (Sometimes, of course, web pages change; the material being analyzed should be saved as screenshots or PDFs to demonstrate what it looked like at the particular time it was downloaded.)

Ethnography and ethics

In any form of ethnography (or netnography), there are ethical issues that need to be addressed. As in covert observation (see Chapter 16), it can become a matter of invading and exposing people's private space. In the case of netnography, it can be argued that the material has already been disclosed in the form of publishing it online. In addition, anonymous aliases are often used. On the other hand, it was disclosed without the intent to allow it to be part of research. It is not always feasible to obtain consent from everyone. User accounts may have been left inactive, for example. Sometimes it may be worth creating pseudonyms and obscuring the identity of participants and forums, but if text is quoted directly, it is often easy to simply google to find out where it was taken from.

Ethical issues must also be addressed in more traditional ethnographic methods. Of course, consent must be obtained, but through conversations and interviews, information that exposes individuals or a third party may emerge. The researcher must then have a strategy for dealing with this. Consulting with one's supervisor is a good option. However, difficult situations may arise, for example, if violations of the law are discovered or if individuals may come to harm. As in other contexts, ethical issues must be taken seriously and dealt with.

***

Ethnography is a method that offers great opportunities to get close to a group of people and to study human societies and cultures. The problem in most thesis contexts is the scope of traditional ethnography. Here, microethnographies and netnographies offer exciting possibilities and make it possible to formulate problems and conduct studies that move away from the all-to-often convenient and non-reflexive choice of interviews.

Ethics is not exclusive to ethnography – such reflection must be part of any research project. I have chosen to discuss it quite explicitly here in the ethnography chapter as it is a method where almost all kinds of ethical issues can occur, simply because ethnography encompasses also other methods. Thus, when it comes to the topic of ethics for other methods: It is no excuse not to consider ethics carefully just because a certain methods book made the pragmatic choice of discussing it in a particular chapter.

PART 4
At the desk

18

WHAT IS AN ANALYSIS?

The analysis is of course a fundamental part of a thesis. It is here that the theoretical framework meets the empirical material, the problem is unraveled, and conclusions are drawn. All through the analysis, the reader should be able to follow the author's train of thought and see how the argument is structured.

It is sometimes debated whether this part of a qualitative thesis should be called 'analysis' or 'interpretation' (see Chapter 3). The type of qualitative method presented here – and for that matter most qualitative methods – is interpretive in nature. However, the part of a thesis discussed in this chapter is usually called analysis, and I stick to that convention, although in many ways it might better be called interpretation.

The analysis will of course be strongly influenced by the theories you are working with. A basic trick when thinking about how to analyze is to go to some other studies that work with the same theories. You don't do this to blatantly copy the approach and style of those studies, but to learn from how they went about doing it, how the analysis is presented, and to think about what might be useful to include in your own thesis. This is another reason not to turn to textbooks when working with theories, as they usually do not account for analytical elements (see Chapter 8).

The craft of analyzing

The very word 'analysis', and especially when combined with 'method' to form 'method of analysis', suggests that there is a clear and systematic approach behind it. Some even argue that another researcher using the same method to analyze the same empirical material would arrive at the same results. At the other end of the scale, we find researchers who argue that qualitative analysis – though here 'interpretation' is the preferred term – is more of an art form than a (traditional) scientific method. Interpretation is argued to be highly dependent on the researcher's personality and background. Often less traditional forms of presentation are used, ranging from ethnographies with a clearly subjective authorial voice to poetry, photography, or video presentations.

DOI: 10.4324/9781003498384-22

Typically, most researchers fall somewhere in between, embracing what might be called a craftsmanship approach. While maintaining a social constructionist stance and an interest in allowing the research material to come through in a nuanced way, most believe that intersubjectivity is a central feature of research and that the research process and the argument developed should be clearly accounted for. It is from this perspective that this book is written (see also Ellingson, 2011). This approach affirms the element of craft and judgment that underpins analysis, but also demands transparency in reporting. It is thus of utmost importance that the reader can easily follow the line of reasoning when conclusions are drawn. Analysis, in the more original sense of the word, is about taking something apart into its components. In many contexts, this is highly relevant, including qualitative and interpretive research. Researchers sometimes speak of 'unpacking' something. In the analysis, the reader can clearly follow how the author has 'unpacked' the empirical material.

An independent stance

In general, analysis is about trying to say something interesting about that empirical material which has been so painstakingly gathered and compiled. Exactly what you say depends on the problem that has been formulated, and this also implies that an analysis demands an independent stance.

The following may sound obvious, but it is something that tends to be forgotten: The empirical material does not speak for itself. It is not enough to have collected fascinating empirical data; it is also important to say something interesting and significant based on that data. The latter is the task of the researcher. The analysis should help us – the other participants of the cocktail party (see Chapter 7) – to understand the way in which the thesis solves the problem it set out to engage with. If we want to shed light on workplace issues in a hospital, we can interview nurses about this, for example. But you cannot straightforwardly use their descriptions of the workplace to answer the research question. Unfortunately, this type of simple reporting is common. The nurses' answers should form the basis for an analysis, but this analysis is done by the researcher and in relation to the theory and the problem formulation. In the analysis, then, the theory is just as present as the empirical material. Silverman (1989, p. 218) formulated this central principle as: 'Avoid treating the actor's point of view as an explanation'. The actor – such as the person you are interviewing – has a unique insight into how they experience their everyday life. But that doesn't mean that everything they say about their everyday life is the only answer (we can experience situations in different ways, and we are not always aware of our prejudice or cognitive limitations). Nor does it mean that they have insight into the theoretical issues or the research problem you are addressing in your thesis.

Another, and partly opposite, mistake is to just compare theory with data. Comparing and analyzing are not the same thing. Comparing can be a first step; perhaps a consulting firm presents itself as a flat and unbureaucratic organization. One can then compare with theoretical ideal types (perhaps Weber's bureaucracy; see Weber, 1914/1968) and see what bureaucratic features still exist. But this can only be a first step. The analysis should not stop at noting that similarities and/or differences exist. It should build on this and ask further questions: Why does the organization look the way it does? Why is it described in a certain way when in fact it has other features? What is the image of the organization held by its management – and by those who work at the 'shop floor'? And so on. Stopping at comparisons often leads to rather trivial conclusions. If comparing in this way is enough to answer the research question, it is likely because the research question is too trivial.

It is therefore the researcher's task to say something independent – and empirically and theoretically well-founded – about the problem that the thesis addresses. The empirical data collected and the theory accounted for and discussed are tools for saying something, but they are not sufficient in themselves.

Deductive, inductive, abductive

Often concepts like 'deductive', 'inductive', and perhaps 'abductive' appear in this context. Putting it simply, a deductive approach starts from clear theoretical notions (hypotheses) that are tested against empirical material. 'Is it the case that …?' An inductive approach, on the other hand, starts from the empirical material (without any theoretical preconception) and builds its conclusions solely on this.

These should be seen as ideal types that are impossible to live up to. Pure deduction with testing of hypotheses (so-called hypothetical-deductive method) becomes problematic in qualitative research because of the process of interpretation itself. It is difficult to argue that interpretation is independent of the researcher, and what is then being tested: The explanatory potential of the hypothesis or the explanatory skills of the interpreter? A purely inductive approach also becomes rather untenable, as it is of course difficult to imagine interpretation taking place entirely without theoretical preunderstanding – even proponents of grounded theory argue that theoretical insights play a central part in the analytical process. For example, Glaser and Strauss (1967) discussed how analysis can draw on personal experience as well as theoretical insights. In practice, some form of 'abduction' is usually practiced (the term comes from the philosopher C. S. Peirce, but see Eco, 1983): An alternation between empirical and theoretical reflection, in which one works with the theory, returns to the empirical material, and contemplates what it might mean in the light of the theory; perhaps discovers new aspects of what is being studied that prompt the modifications and additions to the theoretical framework; and then a renewed theoretical insight meets the empirical material once again. And so on.

It is important to note that abduction is thus not some kind of 'intermediate' variant of deduction and induction, but that it constitutes a logic of its own. Moreover, unlike induction and deduction, which appear to be almost ideal, abduction is very much a practice, even in everyday life. It also involves not only (formal) theory but also extends to preunderstanding and ideology (Alvesson & Sköldberg, 2018).

In the novel *The Name of the Rose*, novelist and semiotician Umberto Eco illustrated this approach through a medieval detective story. A series of murders take place, and the book's 'detective', a Franciscan friar named William of Baskerville, tries to solve the mystery and identify the perpetrator. In doing so, he posits tentative explanations (theories), which are gradually modified and refined in the light of new information (new empirical evidence that emerges as the murders continue, but also through further investigations that are themselves guided by the tentative explanations). Neither the empirical evidence nor the theory is thus set in stone. Rather, it can be seen as two dimensions that come into contact with each other and are continually reshaped. In this way, an explanation is progressively chiseled out. In a similar way, one can work when doing analysis for a thesis in a constant alternation between theory, empirical material, and reflection.

Sort, reduce, argue

But how do you go about it when you're sitting there with piles of printed documents and transcribed interviews on the floor, or a gazillion of files on the computer? How do you know what is important and what is unimportant? What goes with what? And so on. To begin with, some of this diversity should have been dealt with at an initial level, that is, when the problem is formulated and when the methodology is designed. The basis for dealing with a situation where you are drowning in material is to make sure that you do not end up there in the first place (Kvale, 1996).

On a general level, the analytical process can be described in terms of three basic procedures: Sorting, reducing, and arguing (Rennstam & Wästerfors, 2018).

When you sit down with your entire material, you first have to familiarize yourself with it by reading through it several times and starting to reflect on what it's all about. Then you can start sorting it into different categories (this is often called thematizing): One pile about this, one about that, and one about something else. This is a first step and it can be done more or less systematically. In grounded theory, this step is very systematic (Charmaz, 2014; Glaser & Strauss, 1967). The theoretical interest also plays a major role here, together with the researcher's prior understanding of the topic at hand. Sorting is always done with a certain aim (that of the thesis) in mind. It is however important not to decide too early on a definitive thematization. On the contrary, it is often advantageous to try to twist, turn, and 'massage' the material in different ways. Are different categories related to each other? Should one category be seen as a sub-category of another? What makes this piece of information belong to one category and not another? And so on. If you are more than one person working on the thesis, you can take advantage of this and carry out parts of the analysis independently to see what similarities and differences you come up with.

In the next step, you start reducing the material. It is not possible to present everything in the final text. Parts of the empirical material can be condensed. For example, some sequences of events can be described in general terms, while others that are more interesting for the analysis are described in more detail. Of course, it is important that the empirical material is represented in a fair way and that complexities are not glossed over. At the same time, this part of the analysis is strongly driven by the questions to be answered, which means that the selection of material will be strongly dependent on the research questions and aims. There is always a risk that the material is overly reduced and that contradictions and paradoxes (see as follows) are lost, and this is something to be wary of. Often, it is precisely these things that are theoretically interesting.

In the last step, the analysis becomes part of the argumentation of the thesis. Grounded in the problem formulation, the analysis and discussion must clearly lead forward to the conclusions. For example, if the aim is to contribute to conceptual development, the concepts developed must be reflected in both the theory and the empirical material. If the aim is to shed light on a new type of empirical phenomenon using an existing theory, then different aspects of the phenomenon must be highlighted and you ideally also show how under-standing this phenomenon affects the way we relate to the theory in question. If the aim is to demonstrate a theoretical paradox, then both sides of this paradox must be highlighted, and data can play a prominent role in detailing how actors actually deal with or construct the contradictions that we identified. Seen in relation to the writing process, it is also now that the text begins to find its final form. As the text progresses through write-work, the final presentation form starts to take shape. Writing an analysis requires a progressivelyincreasing attention to the way the text will eventually speak to a reader.

Complexity and the dangers of simplification

Empirical material often contains contradictions, paradoxes, and suchlike. An important part of the analysis is to not downplay these. When transcriptions and field notes are turned into an empirical narrative, they are often sorted into themes, as mentioned above. It is then important to be alert to material that contradicts the thematization and suggests that it is not encompassing everything and that it is unambiguous. Different themes can by their contradiction create nuances in the empirical account (Neyland, 2008).

Working with the empirical material involves reading, rereading, taking notes, making markings in the text, discussing with a possible writing partner, sorting, sorting again, rereading ... This phase is often relatively unstructured and exploratory in nature. As mentioned above, you need to work up a familiarity with the material. The familiariza-tion begins as soon as the empirical material is created and does not really end until the thesis as a whole is finished. Creating empirical material, writing, and analyzing is a creative process and in it you need to be nuanced and subtle. Weick's (1989) concept of *disciplined imagination* captures well what it is all about. There must be creativity, but there is also a need to keep it in check. Interpretations must be reasonable and it is important to be able to show the reader clearly both how they emerge in analysis and discussion and how they are relevant to the problem at hand.

A final aspect that should be highlighted is the importance of being nuanced also in the theoretical dimension of the analysis. This may seem obvious, but it is not uncommon to see authors who fall into oversimplifications and tend to want to explain everything from a single starting point. This tendency becomes extremely evident in cases where the author is passionate about a certain theoretical angle, and often there are political or other ideological reasons for this. Theories often have a homogenizing effect on thinking. As the saying goes, if your only tool is a hammer, you might end up treating every problem as a nail. That is: If you choose to rely too much on a single theory, there is a risk that the analysis will become too one-sided and heavily dependent on that particular theory. If you have really made up your mind that everything that happens in an organization can be explained by the 'economic man', or by the hegemony of patriarchy, then most things will probably be explained by that particular theory. The analysis runs the risk of being one-eyed and not adding much that is new, but merely confirming what the theories are based on. Empirical curiosity is unfortunately often stifled by theoretical or political preconceptions.

***

Since analysis is about interpretations and reductions of empirical material, there is a great risk that the theories take charge. There are strong reasons to encourage two things here. First, to be empirically curious, sensitive, and nuanced, trying to best represent the entirety of one's empirical material in the analysis. Empirical complexities and paradoxes should be present in the analysis and conclusions. Second, a broader theoretical repertoire helps. Ideally, when writing a thesis, you will have acquired this through your previous education, but even when working with the theoretical framework of the thesis, it is important to maintain an openness to new ideas and approaches. In addition, taking an abductive view, you should return to theory repeatedly to get new ideas and perspectives. In this way, a qualitative analysis can be interesting, insightful, nuanced, and respond well to the research problem you formulated.

19

WHAT IS CRITIQUE?

One of the most difficult parts of the thesis process is critique. At the same time, critique has a very central function: It is a powerful tool for improving a text. But why is this so difficult?

Critique is not only about giving critique, but also about being able to receive it in a good way. As a thesis writer, you will receive critique from supervisors and often also from your peers. The question of how to deal with critique is central and it is important that those giving the critique can feel confident that it will be received in a good way. Hopefully, no one wants to hurt anyone with their critique.

There's critique and then there's critique

Social science is inherently critical – at least according to a number of thinkers in the field. German philosopher Jürgen Habermas (1968/1987), for example, argued that there are three main knowledge interests. The *technical* is directed toward trying to achieve control over systems, for example in physics, medicine, or biology. The *practical* knowledge interest concerns mainly the historical and human sciences that aim to achieve mutual understanding of the human condition. Finally, the *emancipatory* or *critical* knowledge interest, of particular significance in the social sciences, aims to free us from misconceptions and ideology and to unveil power relationships. Obviously, we cannot draw sharp lines between these, but Habermas' argument indicates that the critical element is quite indispensable to social science.

Following this, critique and critical thinking in this sense should be a mainstay of social science. How the social sciences are doing in this respect is debatable, and it can be argued that many social scientists could be a little more critical and skeptical about the state of things in the area they are studying. What automatically comes next is, of course, a wide-ranging discussion about what this 'being critical' is all about and how it relates to different kinds of values and ideologies. This is, in turn, a question that is related to different theoretical and epistemological positions (such as Habermas' above) and therefore falls outside the scope of this book. For the purposes at hand, it is enough to

DOI: 10.4324/9781003498384-23

note that critique in a Habermasian sense is for many the essence of social science – and for others, it's a red rag.

But there's also another dimension of critique that I will deal with more extensively here. That is the element of critique built into the very organization of science where studies, knowledge claims, and conclusions are constantly under scrutiny. Sociologist Robert K. Merton (1942/1996) described science as 'organized skepticism'. It is the foundations of such skepticism, at a very hands-on level, that is the theme of this chapter. This is critique in a much more limited, but no less important, sense than that which Habermas discussed. This chapter is about how we can practically work with critique when writing an academic thesis: About what constructive critique is, about reading critically, and about how we can productively engage with the critique we receive.

Constructive critique?

Sometimes critique becomes pointless and vague, and that doesn't really benefit anyone. Sometimes too much focus on so-called 'constructive critique' leads to critique not addressing shortcomings and problems just because it might make the writer sad or disheartened. Sometimes 'constructive critique' means that the critic is expected to not only deliver critique but also suggest how to solve a potential problem.

The question is, however, what good such ways of approaching critique do. After all, not voicing critique in order not to upset anyone only leads to that critique coming later or to the end result being less good than it could have been. Demanding that the critic should be able to come up with solutions to all problems that they find is not relevant in all situations. Often, pointing out shortcomings can lead to a discussion which in turn leads to solutions, even if these were not obvious at the outset. Just because you see a problem does not mean you have to have a solution. Methods such as 'two stars and a wish' may be nice and appropriate in primary education, but at higher academic levels, they run the risk of undermining the opportunity for substantive critique.

I think that the problems and concerns that critique often generates can be traced back to critique being viewed in an unproductive way. We must understand that critique is about how a text resonates with a reader; it is not about the competence of the author. This applies, of course, to the critic as well as the recipient of critique. Although it can be difficult to embrace this approach all the time, especially when a lot of time and effort has been put into writing a text, a lot is gained if you can do it. The academic seminar is also – or at least should be – a forgiving environment where we can play around and experiment with different ideas as well as explore their relevance. The seminar is the laboratory of the social sciences. In the lab, some experiments succeed, others don't, and the same is true when we test ideas and arguments in a seminar.

Emic and etic critique

As a critic, when you read a text, you have to think about what in the text it is that you are criticizing. There are two main types of reading here: Let's call them emic and etic critique.

Emic critique is about reading a text 'from the inside', focusing on the logic of the text itself. Do the different parts of the argument hang together? Can you see at each step how it moves, from A to B to C? How does the overall logic work: Does the problematization follow from the problem background? Is the aim relevant? Are the research questions answered? Is the aim fulfilled? Here, the cross-relationships between the thesis' components become key. (See Figure 6.2.)

This is also about assessing the resources that are drawn in to meet the aim. Is the method relevant for the problem at hand? Is there a convincing argument that the method is appropriate and that weaknesses are dealt with? Which theories have been chosen, and why? Is it clear why these particular theoretical choices were made? In short, emic critique is about whether or not the thesis comprises a good argument; it is by emic critique that the craftsmanship is scrutinized. In emic critique, the reader does not care whether or not they agree with what the thesis arrives at, only whether it arrives at this in a reasonable way, grounded in the starting premises. This means that emic critique is fundamentally independent of the critic's own position.

The *etic* critique, on the other hand, comes from 'the outside' and concerns exactly those starting points or premises. It discusses why a particular question is interesting and what the answer that the thesis comes up with might mean in a broader context. For an etic critique to be interesting to engage with at all, the thesis must hold up (reasonably well) to the emic critique. After all, if the thesis's arguments themselves don't hold up, there's little point in discussing the conclusions, since it basically means the conclusions are poorly supported. The etic critique puts the thesis in a wider context: The scientific context, but also the social and political. A thesis in the social sciences is part, albeit a small part, of the ongoing discussion about how we understand and shape society. With this in mind, etic critique is an important dimension (and of course, as a researcher, you should have this in mind from the outset). Etic critique thus approaches what I discussed above in terms of the critical nature of social science and the different critical dimensions are thus in some ways interrelated.

Doubting and believing

Another idea has been developed by Peter Elbow (see for example Elbow, 2008) and involves supplementing the usual scientific skepticism and doubt with a reading style based on wanting to believe. Elbow contrasts the doubting game with the believing game.

Reading a text and engaging in *the doubting game* means reading as a skeptic or as the devil's advocate. One approaches the text with an extremely questioning and critical attitude. Every bit of the argument is scrutinized and one reflects on whether it really works. One reads with the ambition to find errors and flaws, contradictions, and weaknesses. Engaging in the doubting game means being a bit of a cynic, taking nothing in the text for granted.

The opposite is to engage in *the believing game* as an advocate of the text. The advocate wants to agree with the text and wants to be convinced of what it says. One sympathizes with the starting premises of the text and tries to truly understand what this position means. The reader puts themselves in the author's shoes with the ambition to understand their starting points and reasoning as best they can. In this approach, the reader draws on the strengths of the text and tries to find support for what it says.

Perhaps it is mainly the doubting game that we associate with scientific review processes and critique in general. It is undoubtedly a central part of research practice and an important task for a scientific seminar. The point, however, is that if the doubting game is combined with the believing game, the critique becomes more multifaceted and productive. When the two reading styles are combined, one can see the shortcomings of the overall ambitions of the text. A reader who has genuinely tried to understand and believe in the argument a text is making can also more easily see what parts are missing for the text as a whole to work: What is it that the text wants to convey here, but doesn't quite get across? The risk of working purely with doubt is that critique focuses on flaws (which is good) but misses opportunities, that is, things that could strengthen the text but are not currently there. (There are also important learnings for the reader in engaging in the believing game. It forces you to challenge your own assumptions and see things from other points of view. This is important, but falls a little short of the theme of this chapter.)

Both these reading styles thus have their merits, especially in combination, and both can be applied when critically engaging with a text. It is often useful to start with a reading based on doubt and then do one based on believing. It can be difficult to do them simultaneously. By comparing the outcome of the two readings, it is easier to make a synthesis of the strengths and shortcomings of the text.

One should not confuse these 'games' with an emic or etic reading (as per above). While the emic/etic distinction is about which aspects of the argument one looks at, the belief/doubt distinction is about the approach one takes as a reader. Combining the two dimensions gives us four possibilities: We can make an etic reading based on both belief and doubt, and we can make an emic reading based on both belief and doubt.

Precision and relevance

One element that must not be forgotten is that the critique must be relevant and precise. For example, critique of the way an argument is constructed should be illustrated by showing where in the argument the reader does not follow, what gaps there are, and what parts seem to point in the wrong direction. By anchoring also more general critique in examples from the text, it becomes easier to relate to. The recipient of the critique is faced with the task of decoding the critique: What does this mean for the text? How should the text be revised so that the points are more clearly understood by the reader?

Perhaps it should not be necessary to point out that the critic does not resort to what could be experienced as personal attacks and such. Yet it is easy, especially in etic critique, to end up in that direction. Etic critique approaches questions of values and ideology, and discussions of such should always be conducted with respect for the person who is expected to receive the critique.

Receiving critique

The above is mainly about the attitude of the reader, the critic. As a recipient of critique, being aware of these different reading styles is good as it helps in the interpretation of the critique. Developing some distance in the way in which one engages with critique is good, but it is often easier said than done.

In an article entitled 'Turning lemons into lemonade' (2003), organizational scholar William Starbuck presented a productive approach to receiving critique. When you receive harsh and sour critique, it can be easy to feel that you have failed and that what you thought or did or wrote was wrong. There is then a risk that the thesis process will grind to a halt, that you will feel that you have to do it all over again and perhaps would prefer to throw in the towel, feeling that perhaps you should be doing something completely different with your life. Starbuck's idea, however, is to reverse what critique means: Critique is not primarily about the text or the author of the text, but about how the text resonates with an intended reader.

Let's say that a reader of the text points out that an argument doesn't hold together (a fairly common form of emic critique based on the doubting game). One way to understand such a comment is to imagine that you are completely wrong, possibly slightly stupid, and that the argument needs to be completely scrapped, or at least reworked from scratch. But another way of looking at it is to approach it thus: This reader – who represents all potential readers – has not been convinced that the argument makes sense as presented in the text at the moment. Instead of seeing it as a critique of your ability and capacity, you see the critique as helping you to communicate more effectively what you want to say in the text and to reach the relevant audience more easily with your message. The critique will therefore concern the interaction between the text and the reader, not the producer of the text. The key dimension of critique is, then, that you have learned something about the reader. You can ask yourself 'Why did the reader react this way?' and then think about whether you want that reaction. If not, what should you do to get the reaction you want?

Regardless of which way you relate to critique, the consequences are quite similar. In both instances, the text must be revised. But it's easier and more productive to think of it as an act of communicating something to someone in a convincing way than to think of it as dealing with the consequences of your own shortcomings. This does not mean that one should shrug off critique and, so to speak, blame the reader. Remember that the author is responsible for their text (see Chapter 9). Critique should be taken seriously, and a thesis that can withstand a qualified reading based on the doubting game has the potential to be a good thesis. With a bit more distance to critique, harsh critique also becomes good critique – sour lemons turn into sweet lemonade. The responsibility for 'constructive critique' thus lies as much on the receiver as on the sender: It is about being able to receive critique in a constructive way.

***

In the introduction to the book, I mentioned that method is not just about something written in methodology chapters, nor is it something that only exists in theses or inside of the walls of the university. In discussing the functions of critique, this becomes particularly clear. In different contexts, we are expected to be critical and to be able to receive critique, and in order to be able to do so productively, we need to develop strategies and attitudes to do so. The approach to critique I want to advocate in this book is not about being skeptical or doubtful in general, but it is about trying to look seriously and systematically at how questions, arguments, and conclusions are presented and substantiated. This lays the foundations for a substantial and initiated critique (rather than a 'post-truth' attitude; see McIntyre, 2018), a critique that can help to develop a conversation and advance reasoning in many areas of life.

20

WHAT IS QUALITY?

So, here you are: With a well-chiseled-out research question, an empirical study, a theoretical framework, an analysis, and hopefully some insightful conclusions. But how can you know how good those conclusions are? How can you be sure that the study has come up with some kind of relevant knowledge? More generally, what criteria can be used to assess the quality of qualitative research? If we give up the idea that we can simply talk about a correspondence between theory and data – does that mean that 'anything goes', as famously put by philosopher Paul Feyerabend (1975/1993)?

These questions are by no means simple and in many ways concern what research is all about. The quality of research must be judged against questions of what knowledge is and how it is created (epistemology) and how reality is constituted (ontology). The philosophy of science underpinning a research field will inevitably be part of how the concept of quality is approached within that field. For example, Roulston (2010) shows how quality in interview contexts takes on different meanings depending on philosophical positions. This short book does not have space for a more extensive debate on the philosophy of science, but in the following, some concepts and considerations that are commonly used in discussions of quality in qualitative research are introduced. It is also important to put this in relation to the theoretical 'home base' of the thesis; different research fields have developed different traditions of what is considered 'good science'.

The problem with reliability and validity

A common way of discussing quality in scientific contexts is to make a distinction between reliability and validity. These concepts are taken from research based on the correspondence theory of truth (see Chapter 4), and therefore, a traditional use of them is rather ill-suited to qualitative interpretive research.

Reliability refers to whether research outcomes are replicable. Can we trust the measurement? If we do the same study again, will we get the same results? Ideally, another researcher should be able to do this. In many contexts, we see accounts of where several independent studies using the same measurement tools have produced the same results,

DOI: 10.4324/9781003498384-24

which suggests high reliability. Validity, in turn, refers to whether we are actually studying what we want to study. Did what we could measure really measure what we wanted to measure? We can therefore have high reliability without having high validity. If we send out questionnaires to find out how competent car drivers are, and ask questions about this, we might on repeated occasions get the answer that about 90% of all drivers consider themselves to be among the best 50%. The repeated results indicate high reliability, but after some reflection, we realize that the validity is probably quite low.

The correspondence theory of truth holds that we can achieve a correspondence (consistency) between theory and empirical evidence, between the map and reality. But if we imagine that the very process of drawing the map creates reality, that our language and concepts influence our very understanding of reality, this correspondence becomes rather pointless – it will occur automatically.

The problem is that the correspondence approach to reliability and validity is rarely in line with what qualitative interpretive research is all about. If we look at the underlying assumptions and ambitions of many types of qualitative approaches, it is rather meaningless to talk about a reality that is independent of the attempts to describe or conceptualize it. This makes the pair reliability/validity a bit strange, since it is based on the idea that measurements and instruments exist independently of what is being measured.

Interpretive processes are based on actively selecting elements of reality and placing them in a specific context, from observing and writing down our impressions from the field to presenting the material to the reader, and it is quite pointless to claim that 'reality' can be decoupled from either theoretical preconceptions or the role of the interpreter. The simple correspondence theory does simply not capture this very well. This goes for quantitative approaches too, the critics argue, as they also fundamentally rely on conceptualization in the same way. From an interpretive point of view, something such as the claim of replicability becomes rather odd, since the researcher is an active participant in the various stages of the process. It is difficult to imagine that two interviewers will get exactly the same answers if they interview the same person, or even that the same person will deliver exactly the same statements in two interviews that will take place at different times. And in the analysis and interpretive work, where theories and empirical material are related to each other, the interpreter themself naturally plays a central role.

Interpretation, and thus the role of the interpreter, is of course not a problem but a prerequisite. All this has led to other suggestions on how to discuss the quality of qualitative research. In the following, I will introduce three ways of approaching validity in the hope of helping to liberate qualitative researchers from less valid approaches to validity.

Validity I: From craftsmanship validity to pragmatic validity

One way of approaching the concept of quality in qualitative research is to focus on the practical usefulness of the research results. Kvale (1995) discusses how the idea of validity can be reformulated in qualitative contexts. Instead of focusing on correspondence and measurement methods, Kvale reflects on how truthful research outcomes can be substantiated. He presents three types of validity, all of which have relevance. Moreover, they can be seen as building on each other: Craftsmanship validity is a prerequisite for communicative validity, which in turn is a prerequisite for pragmatic validity.

The first is *craftsmanship validity*. This means that results and conclusions are based on systematic work in data collection and analysis. By continuously reviewing the analysis and data, the quality of the reasoning is questioned and checked. It also involves using methodological insights to build an argument for one's empirical and analytical approach. Finally, craftsmanship validity is based on the problematization of the phenomenon at hand, trying to find new and interesting approaches and potential for generating insight. Craftsmanship validity thus permeates all the different parts of a thesis, from problematization through analysis to discussion and conclusions. It also presupposes transparency, as the reader of a thesis must be able to follow how the line of reasoning is built up and how the empirical material was constructed.

The second form is *communicative validity*. Here, the knowledge claims that are made are tested in dialog, where the relevance of the reasoning and outcome is examined. This may be in dialog with the group being researched, with the research community, or with the general public. Here, of course, questions of how the dialog should take place, who should be involved, and, not least, how power relations should be managed come into play. There is a risk, for example, that research findings that are anomalous are marginalized when there are strong established voices within a field (see Dreger, 2016 for illustrative examples). Generally, there is of course a key point here in that knowledge must be communicated to someone in one way or another in order to have any sort of social relevance.

The third and final aspect that Kvale highlights is *pragmatic validity*. This is about the relevance of knowledge in that it can be used to influence society. As famously put by Marx (1845): 'Philosophers have hitherto only *interpreted* the world in various ways; the point is to *change* it', or in Kurt Lewin's (1951) adage that there is nothing as practical as a good theory. As with communicative validity, questions of *For whom?* come into play here. For example, is it about usefulness to the group being researched or to the scientific community? Different types of research results become relevant for the two. Furthermore, the question of power is raised again. Research here becomes explicitly normative and there is a risk that the empirical curiosity and knowledge-creation become overshadowed by a political agenda, whatever that agenda might be. There is thus a serious risk that research becomes merely a way of dressing up a predetermined opinion in fancy words and giving it legitimacy through a scientific label.

However, the counter-argument here has its relevance. Some argue that knowing always has a political dimension, and it is more a question of making this visible than hiding behind supposed objectivity. From a pragmatic criterion of validity, then, the results of a study will be about action implications: 'Knowledge is action rather than observation; the effectiveness of our knowledge beliefs is demonstrated by the effectiveness of our action' (Kvale, 1995, p. 32). However, care must be taken to ensure that pragmatic validity does not turn into demagogic argumentation or that the ends are allowed to justify the means, so that the other validity criteria are set aside in favor of pragmatic validity. In particular, craftsmanship validity should serve as a safeguard against this. Also, what was referred to in Chapter 7 as the actor approach to problem formulation is a way of dealing more reflexively with these types of issues.

Validity II: Pointfulness

The Swedish sociologist Johan Asplund, in a 1970 book that is still relevant today (but unfortunately not available in English), put forward a very different criterion for the quality of a theory: Pointfulness. The term indicates insight, pertinence, and meaningfulness. His idea is that social science must for sure deal with various forms of verification but that underneath this there is always a form of insight that is not in itself verifiable. We cannot empirically determine or verify whether the idea of society as an organism or a machine is a better image than, say, the idea of society as a contract. We can only say whether one image is more pertinent and insightful than the other in a certain context for a certain purpose. Any social theory of relevance always has an element of innovative pointfulness to them. Pointfulness is characterized by an 'Aha!' experience: 'Now I see!' Some might argue that discussing this in terms of validity is taking it a step too far, but I would still argue that the idea of pointfulness provides an interesting and challenging way to discuss quality.

Asplund's idea of pointfulness is at once modest and pretentious. It leaves the relationship between theoretically relevant – or perhaps rather interesting – insights and empirical work fairly open. Empirical work is important and central and can in itself give rise to pointful insights. But it is, to play with words, the pointfulness that is the point, not the theory or the data in and of themselves. In a way, this makes research relatively simple – it's 'only' about coming up with new and interesting ideas. At the same time, this is a really tough criterion. It is not easy to come up with insights that are new and truly insightful and pertinent. Often, a sense of novelty mainly indicates a lack of insight into the history of a topic or theoretical area. Notably, Asplund did not argue that all research should be about producing revolutionary insights – that would be almost paradoxical: If new insights were produced all the time, each one would hold relatively little value. On the other hand, there are good reasons to give value to insights that help us to reformulate old problems in new ways and to see new aspects of what we thought we already knew, even if this cannot be verified empirically by traditional means. Seeing something is a prerequisite for being able to at all discuss it, for example, in terms of correlation or correspondence. Pointfulness thus becomes an overriding criterion for the quality of a theoretical contribution.

Validity III: Authenticity

Another form of validity that is sometimes put forward is that the results should be authentic (cf. Lincoln & Guba, 2000) and that those studied should be able to recognize themselves in the descriptions. In studies of cultures, for example, this is often stressed as a key criterion: Both those being part of the culture and those confronting it from the outside must feel that the description is authentic. This sounds reasonable, of course, but at the same time, it is important not to get caught looking for the lowest common denominator (Kvale, 1996). If everyone has to identify with a description, it means that voices on the margins will be exactly that: Marginalized, and thus remain marginalized. Claims that those studied should be able to recognize themselves in descriptions must be taken with a truckload of salt. Otherwise, there is a risk that findings highlighting contradictions and paradoxes (see Chapter 18) will be silenced and that the desired nuanced and multifaceted description of social reality will become one-dimensional. This becomes even more accentuated in critical research that explicitly challenges and questions for example power relationships. Of course, there may be individuals who do not identify with descriptions or even find themselves uncomfortable with them – but why should they have the last word? After all, the point of such research is often precisely to question what we thought we knew. (See further the section 'There's critique and then there's critique' in Chapter 19.)

For example, sometimes researchers try to increase the validity of interviews by sending the transcribed interviews back to the respondents to let them ensure that what was said was expressed and understood correctly. This is said to increase authenticity by giving the respondent a second chance to clarify. This can sometimes be a good thing, and the respondent can have the chance to add things they forgot to say during the interview or that they thought of later. However, it can also have other, less favorable, consequences. I once interviewed a manager about a change process he was responsible for. When asked during the interview how he thought about how to structure such a process, he replied that 'uh, well, you know, you have to stir the pot a little bit now and then, make things happen'. When the transcript of the interview came back after the respondent had read it, that comment had been deleted and replaced with the text: 'When implementing change, it is important to have a clear process of creating legitimacy in the organization, where ...' Blah, blah, blah; a typical pop-management response followed, describing with textbook-like pretentiousness a clinical and anemic view of implementation processes. The question is, of course, whether this was much more authentic than his spontaneous response about stirring the pot ...

Thus, authenticity is not a simple criterion. As in all other cases of qualitative research quality, it is about making judgments. It is partly about you continuously assessing the quality in the writing and analysis processes, but also about writing your thesis in such a way that it can be assessed by a reader. Being assessable and transparent, however, does not mean adapting to the reader and going fully along with their expectations. On the contrary, good research can challenge the reader's assumptions.

Empirical and theoretical generalization

The ambition of a study is often to be able to say something about a phenomenon in general, that is, to generalize. Here, however, it is important to distinguish between an empirical and a theoretical generalization. An *empirical generalization* involves making statements that are valid for all instances of the same category based on a limited number of instances studied – this may be a set of interviews or case studies. An example might be that you conduct a number of interviews with divorced parents who live every other week with their children and then want to make a general statement about what it is like to be an every-other-week parent. In this type of generalization, you will have to deal with issues of sampling and representativeness of the material studied – issues that can be favorably addressed with traditional concepts of validity and reliability (see also Chapter 11).

However, this is not the only type of generalization that is possible to make; there is also *theoretical generalization*. This approach is not about saying that phenomenon x looks like y in z% of all cases. Rather it is about showing how the theory or concepts one has developed can be used to understand at least one, and possibly more, instances of a phenomenon. Say you want to study how strategic branding takes place in public organizations, focusing on how a municipality works on its brand. Based on that study, you cannot make a statement about how municipalities in general work strategically with branding (some municipalities may not even care about their brand), but you can understand how conditions specific to this type of organization (e.g. governance based on representative democracy) may affect practices that have traditionally belonged to the domains of the for-profit sector (e.g. brand positioning). Building on your case, you may be able to suggest concepts for understanding such processes, which can then enrich the understanding of other cases of the same category (public organizations working with branding). We can here talk about extending the application of a theory to new cases within the domain in which the theory is potentially relevant. In the case just imagined, the theory of strategic branding may potentially apply to all organizations operating in some form of market; the current study explores in more detail what a particular type of case (a public organization) may look like and what changes to the theory might be needed in order to make it work in that case (when 'market' may not be a fully adequate term to describe the organization's environment). What the reader takes away from the thesis is not a general statement about how branding strategies are always or necessarily designed in public organizations, but a refined conceptual apparatus that can be used to understand and discuss other such cases.

There is therefore a difference between different types of generalizations, and this in turn means that there is a difference in the type of criteria that the study must meet in order to be said to be of good quality. Concepts of validity and reliability derived from the correspondence theory of truth are usually of marginal relevance in interpretive research, since the claims made are normally not about empirical generalization. Theoretical generalization opens up other ways of assessing validity, such as the three variants discussed above.

The necessary transparency

A common theme in discussions of research quality, that I have mentioned several times in this book, is *transparency*: That the research can be reviewed and that all relevant data and all steps in the line of reasoning are presented. To some extent, this becomes difficult in qualitative research. One problem is that not all data can usually be reported in a straightforward manner. In any account of research, such as a thesis, only a small proportion of all the interview material, field notes, and so on will be represented (see Chapter 18). Even if, for example, interview transcripts and field notes can be made available to a reviewer or an opponent – which in turn is complicated if anonymity has been promised – this does not mean that all the material can be made available. If you have done field studies with observations, then you should allow yourself to be influenced by the impressions you get during the field work, impressions that cannot always be easily expressed in words. In most cases, then, we have to rely on the researcher's honesty and on the fact that they have allowed the material to be represented in a fair way and that they have not been cherry-picking data.

The transparency of the argument, the written text, is easier to deal with. For this, it is a matter of presenting your argument in such a way that a reader can follow and understand it. Critique, as discussed in the last chapter, is an important part of the writing process and, if done well, gives good indications of how the text resonates with a reader in this regard. Learning how to structure the argument in a clear way, such as building the thesis around the U-model, is important in giving the reader a good chance to examine the argument. In general, then, transparency in accounting for data and in textual presentation is a criterion of quality that cannot be compromised – it is a prerequisite for assessing quality.

The various quality criteria discussed above – craftsmanship validity, communicative validity, pragmatic validity, pointfulness, and authenticity – may all be important. Transparency, however, is a prerequisite for these quality criteria to even be considered. Transparency is a *sine qua non*: Without it, nothing. As in many other contexts, when it comes to qualitative research, it is a matter of giving the reader the chance to make an informed judgment about whether a thesis holds good quality.

***

In the case of assessing quality, the broader value of method – for an academic education but also for developing the necessary skills as an informed citizen of society – comes into play again. We are often confronted with situations that require our own judgment and also assessing the judgment of others. Developing methodological skills and a critical gaze is about being able to make such judgments and assessments in a more qualified and nuanced way. This in turn requires an understanding of how questions of quality, in order to be relevant, need to be asked differently in different contexts in relationship to different knowledge claims. By learning this, we learn to make more informed judgments about knowledge claims in general, not just in scientific texts.

21

CONCLUSION

In the introduction to this book, I stated the aim of the book: To introduce, in a clear and concise way, a series of questions that you as a beginner might ask yourself when writing your thesis. Then, in 19 chapters, various issues have been discussed, ranging from what a thesis is to how to conduct interviews and observations, what data and theory are, and how to deal with critique and issues of quality.

As pointed out in several instances in the book, this is only the beginning. If you have read the book as part of an introductory course, this may be the first method book you have ever read (but hopefully not the last). Perhaps you have read it as one of several methodology books in a methods course or as a guide to writing your thesis. The further in your education you have come, the more weight other literature is likely to carry. However, with this book, I have aimed to contribute something unique: To provide a quick overview and to put a number of key issues into their respective contexts.

The starting point of the book is that methods in qualitative and interpretive research cannot be reduced to a set of step-by-step techniques that should be followed mechanically. Rather, it is a combination of skills that one must learn to master, and it involves making informed judgments and trade-offs.

The craftsmanship of method

Sometimes research is presented as a question of craftsmanship. By this is meant that it is not something that can easily be done by following detailed procedures, models, or recipes. Rather, it is something that is developed through participation and in which one becomes progressively involved. In the beginning, you may write short papers focusing on formulating research problems and mastering citations; in methods courses, you will perhaps practice some specific techniques; and in advanced courses, you will practice the ability to handle more advanced theory and to create independent arguments based on theory (rather than making exam-style write-ups; the latter also play a role in education, but a different one). Finally, it all culminates in a thesis, where you demonstrate that you

DOI: 10.4324/9781003498384-25

have mastered the ability to develop an independent, theoretically informed argument that in some way contributes to knowledge. At all stages, the idea of viewing a thesis as an argument is central to the development of a reflective and analytical approach – the 'critical approach' so often referred to.

Critique also plays a key role in this craft. By reading your peers' work and discussing scientific texts – preferably texts of really high quality, such as classics in a field of research – a critical gaze and the ability to see the strengths as well as the weaknesses of an argument are developed. This is where universities and higher education institutions play a key role in the development of democratic dialog in society. The importance of critique combined with the insight that relevant critique is always based on knowledge is something I like to emphasize both in this book and in my own teaching.

The craftsmanship approach means that method, writing, and critical thinking are things that you are gradually introduced to. It is not always easy to see why, for example, questions of philosophy of science are central but as an apprentice, sometimes you just have to accept and have some confidence that some things become clear only later. I myself remember my first university course, it was in philosophy of science, and how many of us students didn't really understand why we were taking it. After all, we had applied for a degree in business studies, not philosophy! Many were critical. A couple of years later, however, most were very positive about that initial course. Only then did we realize how central that course had been for our approach to knowledge and learning later in our education. I don't think this would have been remedied by any teacher telling us that it was important (I'm sure they did, and we probably didn't quite care). Some insights you simply have to allow yourself to grow into gradually. In developing craftsmanship, the presence of ambiguity is not always a problem, but a key function (Alvehus, 2022).

Developing judgment

Research is (among other things) a matter of judgment. It concerns judgments about how to assess theories, about how to create problem formulations, about choosing appropriate methods, about sampling, about how to present an argument, about analytical reasoning, about validity, and about the relevance of conclusions. There is yet no machine or algorithm, artificially intelligent or not, that can replace this.

Working with methodology means learning to work with this type of judgment. In the context of a thesis process, you will be repeatedly forced to make such judgments and argue for your choices. Through giving and receiving critique, you will develop your ability to participate in such judgment processes. This professional dimension – making informed and initiated judgments about the quality of the research – is not a problem in qualitative research but a prerequisite for it and a point of it.

Today, in society in general and in workplaces in particular, we see an increasing demand for measurement, control, and management. Everything must be accounted for, preferably in numbers (Power, 1997). Universities are not exempt from this trend (Fleming, 2021). It is important, I would argue, to highlight the fact that not everything can be measured and quantified, but that judgment – qualified, initiated, and critically informed – is a central part of many activities (Alvehus, 2022). Qualitative approaches have important insights to convey here, insights that often require dialog, reflection, and time to come into their own. My hope with this book is that I have been able to contribute, however so small, to a greater understanding of these issues.

***

But now it's your turn. You have a thesis to write. I wish you good luck in this and hope it will be an interesting, appropriately challenging, and rewarding journey!

22

FURTHER READING

As I wrote in the introductory chapter, this is not an in-depth book. It is certainly not enough to answer all the questions that will arise in the course of writing a thesis. Rather, the aim of the book is to provide a bit of guidance on some of the first questions you will face as a thesis writer and to point out some key decisions you have to make. To move forward, you need to read more! In the following, I list some books that you may find useful. As before, the research area in which your thesis is written will have its own methodological discussion to which you must relate, and this of course cannot be covered in the list as follows. The selection consists of more generic literature and is mainly based on things that I have found useful myself or that have been helpful to students I have supervised in the past. The list is, of course, extremely subjective.

General methodological literature

For those who want to broaden their horizons and reflect a little more deeply on the whole idea of the (social) scientific project, there is of course an almost infinite amount of literature; the classics can be lined up. For those who want to get into the contemporary discussion, Naomi Oreske's *Why trust science?* (2019) is a good start and I would also recommend some reflections on the role of expertise (e.g. Collins & Evans, 2007; Eyal, 2019) and discussions on the 'post-truth' condition (McIntyre, 2018). For that extra deep plunge, why not revisit Lyotard's *The Postmodern Condition* (1979/1984), a book that to me seems to have regained relevance in the last few years.

Alvesson and Sköldberg's book *Reflexive Methodology* (2018) is often referred to when it comes to interpretive research, and it provides a good introduction to the field. It is relatively detailed but thus also very rich in content. Silverman (2007) provides a personal and initiated introduction to a range of issues in qualitative research, not least around how to consider what constitutes appropriate empirical material. The five (so far) volumes of *The SAGE Handbook of Qualitative Research* (edited by Denzin and Lincoln; the editions are from 1994, 2000, 2005, 2011, and 2017) reflect fairly well the evolution of the qualitative method over the last third of a century. Each edition contains many

DOI: 10.4324/9781003498384-26

different viewpoints, so it may be worth flipping through all of them to see which chapters may be worth a closer look.

A central concept in the methodological debate is social constructivism (or social constructionism). There is an ocean of literature on this, but Hacking's *The Social Construction of What?* (1999) is a good place to start. A true classic in the field is Berger and Luckmann's *The Social Construction of Reality* (1966). A different take on constructivism is represented by Bruno Latour; see *Reassembling the Social* (2005). Otherwise, the main advice here is to turn to the debate in the academic field in which the thesis is written.

This book has focused a lot on formalities and structures. Of course, that is not the only important thing in research and in writing a thesis; equally important, of course, is creativity and inventiveness. Becker's *Tricks of the Trade* (1998) contains a number of ideas on how to approach concrete research projects to 'twist and turn' empirical data, theory, and concepts. I unfortunately cannot direct the international reader to the work of Johan Asplund, as it has not been translated to English. Much of his work is still thought-provoking and forces the reader to think in new ways. The closest we get is Alvesson and Kärreman (2011) which is a development of some basic themes of Asplund and provides more concrete guidance for new thinking in qualitative research. Alvehus (2019b) provides guidance on how to work with problem formulation in a more creative way, also incorporating ideas from Asplund.

Academic writing

Rienecker and Jørgensen's *The Good Paper* (2018) is a good, if sometimes lengthy, text. An excellent resource for learning the craft of writing is the blog *Inframethodology* (http://blog.cbs.dk/inframethodology/), run by Thomas Basbøll. He is a resident writing consultant at Copenhagen Business School and has developed a way of working with scientific writing from which much can be learned. Among other things, I got the idea of seeing the paragraph as the fundamental part of a scientific text from him, and also the idea of writing one paragraph at a time, focused, with a pomodoro clock.

For those who wish to delve further, I find Helen Sword's *Stylish Academic Writing* (2012) and Deirdre Nansen McCloskey's *Economical Writing* (2019) to be quite indispensable.

Writing has, of course, been discussed by writers of fiction – and they should be familiar with it, I guess. Stephen King's *On Writing* (2000) is a fun place to start.

Specific methods: Interviews, focus groups, observations, and ethnography

A broad overview of interviewing as a research method can be found in Gubrium and Holstein's (2002) handbook. Rapley (2004) offers a concise and nuanced introduction to the topic; not least, the discussion of interviewer neutrality is recommended. Kvale (1996) is a broad and thorough introduction to interview research, covering all steps of an interview study, from initiation to writing. A more skeptical view of interviews and their possibilities can be found in Alvesson (2011). For focus groups, practical introductions can be found in *The Focus Group Kit* (Morgan & Kreuger, 1997). Czarniawska (2007) writes about shadowing and observations and Kozinets (2010) about netnography. The idea of microethnography, as presented here, was developed by Alvehus and Crevani (2022) but Pink and Morgan (2013) should be consulted too.

When it comes to ethnography more generally, for an overview, see Atkinson et al. (2001). Van Maanen's *Tales of the Field* (1988) has become a classic when it comes to writing ethnographies and should be consulted. Another obligatory passage point is Geertz (1973), the chapter on thick descriptions maps out a view on ethnographic work that is all but mandatory to read (although feel free to disagree). For those interested in recent controversies and ethical conundrums, see Alice Goffman's *On the Run* (2014) and the subsequent controversies (e.g. Manning et al., 2016).

Analysis

Different interpretive traditions have different ways of looking at the interpretive process, and within the broad tradition known as hermeneutics, there is a wide range of different directions. Added to this are the insights that came with the impact of post-structuralism and postmodernism on the social science scene in the 1980s. An overview of this can be found in Alvesson and Sköldberg (2018), who also present a version of the abductive method. Silverman (2000, 2007) links arguments about analysis to questions of quality in qualitative method and provides solid and practical advice. See also Silverman (1989) for guidance to avoid some analytical pitfalls. In general, it is difficult to describe in detail how qualitative analysis is done, as it is largely a craft learned by doing, but Rennstam and Wästerfors (2018) provide a good insight into the processes. In so-called grounded theory, there are many more hands-on and systematic methods. For grounded theory, see for example Charmaz (2014) and Strauss and Corbin (1990), of course in conjunction with Glaser and Strauss' (1967) near-canonical and still insightful text.

Sometimes, research seems like detective stories. As the above-mentioned Alvesson and Kärreman (2011) argue, it is actually not far from the truth. Umberto Eco's *The Name of the Rose* (1980/1998) makes for an entertaining introduction and a more advanced discussion can be found in Eco and Sebeok (1983).

Critique, quality, and ethics

For an introduction to the critical dimension of social sciences, the work of Jürgen Habermas is indispensable. The short appendix 'Knowledge and Human Interests: A General Perspective' in Habermas (1968/1987) is a good introduction. The work within the so-called Frankfurt School provided the foundations for this; for an overview, see Jay (1973). Robert K. Merton's work deals with the critical aspect of science as a social structure. Many of his ideas are still useful; see the collection of essays in Merton (1996).

When it comes to different notions of validity and quality, there are of course references in that chapter, but I'd like to re-emphasize the value of Kvale (1995). Moreover, Alvesson and Sandberg's *Re-imagining the Research Process* (2021) provides some food for thought in this regard. For those who feel that the interpretive social sciences, as presented here, seem too arbitrary and fuzzy, please see Firestein (2012, 2016) or, of course, contributions in the sociology of science such as Fleck (1935/1979), Latour and Woolgar (1979), Collins (1992), and Knorr-Cetina (1999) – and of course Oreskes' *Why trust science?* (2019). Put briefly, judgment is always at the heart of any scientific effort.

Research ethics has perhaps not received the role it deserves in this book. Partly this is because, in my view, it is something that is deeply embedded within research practice in specific fields: All come with their own challenges and generic guidelines commonly fail to acknowledge this. Modern research ethics are commonly understood as originating with the Nürnberg trials and the code for medical experiments originating from these, but the area is of course much broader than issues of medical experiments can cover. A good introduction to research ethics for social scientists is the appropriately named *Research Ethics for Social Scientists* by Israel and Hay (2006).

REFERENCES

Alvehus, J. (2017). Clients and cases: Ambiguity and the division of labour in professional service firms. *Baltic Journal of Management*, *12*(4), 408–426.

Alvehus, J. (2019a). Emergent, distributed, and orchestrated: Understanding leadership through frame analysis. *Leadership*, *15*(5), 535–554.

Alvehus, J. (2019b). *Formulating Research Problems*. Lund: Studentlitteratur.

Alvehus, J. (2022). *The Logic of Professionalism: Work and Management in Professional Service Organizations*. Bristol: Bristol University Press.

Alvehus, J., & Crevani, L. (2022). Micro-ethnography: Towards an approach for attending to the multimodality of leadership. *Journal of Change Management*, *22*(3), 231–251.

Alvehus, J., Eklund, S., & Kastberg, G. (2019a). Inhabiting institutions: Shaping the first teacher role in Swedish schools. *Journal of Professions and Organization*, *6*(1), 33–48.

Alvehus, J., Eklund, S., & Kastberg, G. (2019b). Organizing professionalism – New elites, stratification and division of labor. *Public Organization Review*.

Alvehus, J., Eklund, S., & Kastberg, G. (2021). To strengthen or to shatter? On the effects of stratification on professions as systems. *Public Administration*, *99*(2), 371–386.

Alvesson, M. (2003a). Beyond neopositivists, romantics, and localists: A reflexive approach to interviews in organizational research. *Academy of Management Review*, *28*(1), 13–33.

Alvesson, M. (2003b). Methodology for close-up studies: Struggling with closeness and closure. *Higher Education*, *46*, 167–193.

Alvesson, M. (2011). *Interpreting Interviews*. London: SAGE.

Alvesson, M., Gabriel, Y., & Paulsen, R. (2017). *Return to Meaning. A Social Science with Something to Say*. Oxford: Oxford University Press.

Alvesson, M., & Kärreman, D. (2011). *Qualitative Research and Theory Development: Mystery as Method*. Thousand Oaks, CA: SAGE.

Alvesson, M., & Sandberg, J. (2021). *Re-imagining the Research Process: Conventional and Alternative Metaphors*. Los Angeles: SAGE.

Alvesson, M., & Sköldberg, K. (2018). *Reflexive Methodology: New Vistas for Qualitative Research* (3rd ed.). Los Angeles, CA: SAGE.

Asplund, J. (1970). *Om undran inför samhället*. Uppsala: Argos.

Asplund, J. (2002). *Avhandlingens språkdräkt*. Göteborg: Korpen.

Atkinson, P., Coffey, A., Delamont, S., Lofland, J., & Lofland, L. (Eds.). (2001). *Handbook of Ethnography*. London: SAGE.

Basbøll, T. (2023a). Background. *Inframethodology.* https://inframethodology.cbs.dk/?page_id=654

Basbøll, T. (2023b). Theory. *Inframethodology.* https://inframethodology.cbs.dk/?page_id=656

Becker, H. S. (1998). *Tricks of the Trade: How to Think About Your Research While You're Doing It.* Chicago: University of Chicago Press.

Berger, P. L., & Luckmann, T. (1966). *The Social Construction of Reality: A Treatise in the Sociology of Knowledge.* Garden City, NY: Anchor Books.

Bjereld, U., Demker, M., & Hinnfors, J. (1999). *Varför vetenskap?* Lund:Studentlitteratur.

Blumer, H. (1954). What is wrong with social theory? *American Sociological Review, 19*(1), 3–10.

Brunsson, N. (1981). Företagsekonomi – Avbildning eller språkbildning. In N. Brunsson (Ed.), *Företagsekonomi – sanning eller moral? Om det normativa i företagsekonomisk idéutveckling* (pp. 100–112). Lund: Studentlitteratur.

Charmaz, K. (2014). *Constructing Grounded Theory* (2nd ed.). Los Angeles, CA: SAGE.

Collins, H. M. (1992). *Changing Order. Replication and Induction in Scientific Practice.* Chicago: The University of Chicago Press.

Collins, H., & Evans, R. (2007). *Rethinking Expertise.* Chicago: The University of Chicago Press.

Corvellec, H. (Ed.). (2013). *What Is Theory? Answers from the Social and Cultural Sciences.* Stockholm: Liber.

Crampton, J. (1994). Cartography's defining moment: The Peters projection controversy, 1974–1990. *Cartographica, 31*(4), 16–32.

Czarniawska, B. (2007). *Shadowing and Other Techniques for Doing Fieldwork in Modern Societies.* Malmö: Liber.

Czarniawska, B. (2011). *Cyberfactories. How News Agencies Produce News.* Cheltenham, UK: Edward Elgar.

Dingwall, R. (1997). Context and Method in Qualitative Research. In Miller, G. & Dingwall, R. (Eds.), *Accounts, interviews and observations.* (pp. 51–65). London: SAGE.

Dreger, A. (2016). *Galileo's Middle Finger. Heretics, Activists, and One Scholar's Search for Justice.* New York: Penguin.

Eco, U. (1980/1998). *The Name of the Rose.* London: Vintage.

Eco, U. (1983). Horns, hooves, insteps: Some hypotheses on three types of abduction. In U. Eco & T. A. Sebeok (Eds.), *The Sign of Three: Dupin, Holmes, Peirce.* Bloomington: Indianapolis University Press.

Eco, U., & Sebeok, T. A. (1983). *The Sign of Three: Dupin, Holmes, Peirce.* Bloomington: Indianapolis University Press.

Eisenhardt, K. M. (1989). Building theories from case study research. *Academy of Management Review, 14*(4), 532–550.

Elbow, P. (2008, April 2008). *The believing game – methodological believing* Conference on College Composition and Communication, New Orleans.

Ellingson, L. (2011). Analysis and representation across the continuum. In N. K. Denzin & Y. S. Lincoln (Eds.), *The SAGE Handbook of Qualitative Research* (4th ed.). Thousand Oaks, CA: SAGE.

Eyal, G. (2019). *The Crisis of Expertise.* Cambridge: Polity.

Feyerabend, P. (1975/1993). *Against Method: Outline of an Anarchistic Theory of Knowledge.* London: Verso.

Fleck, L. (1935/1979). *Genesis and development of a scientific fact.* Chicago: University of Chicago Press.

Fine, M. (1994). Working the hyphens. Reinventing self and other in qualitative research. In N. K. Denzin & Y. S. Lincoln (Eds.), *Handbook of Qualitative Research.* Thousand Oaks, CA: SAGE.

Firestein, S. (2012). *Ignorance: How It Drives Science.* New York: Oxford University Press.

Firestein, S. (2016). *Failure: Why Science Is so Successful.* Oxford: Oxford University Press.

Fleming, P. (2021). *Dark Academia. How Universities Die.* London: Pluto Press.

Flyvbjerg, B. (2011). Making social science matter. In G. Papanagnou (Ed.), *Social Science and Policy Challenges: Democracy, Values and Capacities* (pp. 25–56). Paris: UNESCO (United Nations Educational, Scientific and Cultural Organization).

Flyvbjerg, B., Landman, T., & Schram, S. (Eds.). (2012). *Real Social Science. Applied Phronesis.* Cambridge: Cambridge University Press.

Foucault, M. (1982). The subject and power. *Critical Inquiry, 8*(4), 777–795.

Foucault, M. (1995). *Discipline & Punish. The Birth of the Prison*. New York: Vintage Books. (1977)

Garfinkel, H. (1967). *Studies in Ethnomethodology*. Cambridge: Polity.

Geertz, C. (1973). *The Interpretation of Cultures*. New York: Basic Books.

Gibbs, A. (1997). *Focus groups* (Social research update 19 (Winter), Issue.

Glaser, B. G., & Strauss, A. L. (1967). *The Discovery of Grounded Theory. Strategies for Qualitative Research*. New Brunswick: Aldine.

Goffman, A. (2014). *On the Run. Fugitive Life in an American City*. New York: Picador.

Goffman, E. (1974). *Frame Analysis. An Essay on the Organization of Experience*. Boston: Northeastern University Press.

Gubrium, J., & Holstein, J. (Eds.). (2002). *Handbook of Interview Research*. London: SAGE.

Habermas, J. (1968/1987). *Knowledge & Human Interests*. Cambridge: Polity Press.

Hacking, I. (1999). *The Social Construction of What?*. Cambridge, Mass: Harvard University Press.

Halkier, B. (2008). *Fokusgrupper* (2nd ed.). Frederiksberg: Samfundslitteratur.

Hemingway, E. (2012). *A Farewell to Arms: The Special Edition*. London: William Heinemann.

Hempel, C. G., & Oppenheim, P. (1948). Studies in the logic of explanation. *Philosophy of Science*, *15*(2), 135–175.

Hill, S. S., Soppelsa, B. F., & West, G. K. (1982). Teaching ESL students to read and write experimental-research papers. *TESOL Quarterly*, *16*(3), 333–347.

Israel, M., & Hay, I. (2006). *Research Ethics for Social Scientists. Between Ethical Conduct and Regulatory Compliance*. London: SAGE.

Janesick, V. J. (2000). The choreography of qualitative research design. In Denzin, N. K. Lincoln, Y. S. (Eds.), *Handbook of Qualitative Research* (2nd ed.), Thousand Oaks: SAGE.

Jay, M. (1973). *The Dialectical Imagination. A History of the Frankfurt School and the Institute of Social Research 1923–1950*. Berkeley: University of California Press.

King, S. (2000). *On Writing. A Memoir of the Craft*. New York: Scribner.

Knorr-Cetina, K. (1999). *Epistemic Cultures: How the Sciences Make Knowledge*. Cambridge, MA: Harvard University Press.

Korn, J. H. (1997). *Illusions of Reality: A History of Deception in Social Psychology*. New York: State University of New York Press.

Kozinets, R. V. (2010). *Netnography. Doing Ethnographic Research Online*. London: SAGE.

Krueger, R. A., & Casey, M. A. (2008). *Focus Groups*. Thousand Oaks, CA: SAGE.

Kvale, S. (1995). The social construction of validity. *Qualitative Inquiry*, *1*(1), 19–40. 10.1177/107780049500100103

Kvale, S. (1996). *InterViews. An Introduction to Qualitative Research Interviewing*. Thousand Oaks, CA: SAGE.

Lakoff, G., & Johnson, M. (1980). *Metaphors We Live By*. Chicago: The University of Chicago Press.

Larsson, M., & Alvehus, J. (2023). Blackboxing leadership: Methodological practices leading to manager-centrism. *Leadership*, *19*(1), 85–97.

Latour, B. (2005). *Reassembling the Social. An Introduction to Actor-Network-Theory*. Oxford: Oxford University Press.

Latour, B., & Woolgar, S. (1979). *Laboratory Life: The Social Construction of Scientific Facts*. Beverly Hills: SAGE.

Lekvall, P., & Wahlbin, C. (2001). *Information för marknadsföringsbeslut* (3rd ed.). Göteborg: IHM. (1979)

Lewin, K. (1951). *Field Theory in Social Science: Selected Theoretical Papers* (D. Cartwright, Ed.). New York: Harper & Brothers.

Lincoln, Y. S., & Guba, E. G. (2000). Paradigmatic controversies, contradictions, and emerging confluences. In N. K. Denzin & Y. S. Lincoln (Eds.), *Handbook of Qualitative Research* (2nd ed.). Thousand Oaks, CA: SAGE.

Lofland, J., Snow, D. A., Anderson, L., & Lofland, L. H. (2006). *Analyzing Social Settings. A Guide to Qualitative Observation and Analysis* (4th ed.). Balmont, CA: Wadsworth/Thomson Learning.

Lyotard, J.-F. (1979/1984). *The Postmodern Condition. A Report on Knowledge (Vol. 10)*. Minneapolis: The University of Minnesota Press.

Malinowski, B. (1922). *Argonauts of the Western Pacific: An Account of Native Enterprise and Adventure in the Archipelagoes of Melanesian New Guinea*. London: Routledge & Kegan Paul.

Malmsten, B. (2012). *Så gör jag: Konsten att skriva*. Stockholm: Modernista.

Manning, P., Jammal, S., & Shimola, B. (2016). Ethnography on trial. *Society*, *53*(4), 444–452.

Marx, K. (1845). Theses on Feuerbach. Retrieved 2018-09-25, from https://www.marxists.org/archive/marx/works/1845/theses/

McCloskey, D. N. (2019). *Economical Writing. Thirty-Five Rules for Clear and Persuasive Prose* (3rd ed.). Chicago: The University of Chicago Press.

McIntyre, L. (2018). *Post-Truth*. Cambridge, MA: The MIT Press.

Merton, R. K. (1942/1996). The ethos of science. In R. K. Merton (Ed.), *On Social Structure and Science*. Chicago: The University of Chicago Press.

Merton, R. K. (1996). *On Social Structure and Science*. Chicago: The University of Chicago Press.

Milgram, S. (1975/2005). *Obedience to Authority: An Experimental View*. London: Pinter and Martin. (1974)

Mol, A. (2002). *The Body Multiple: Ontology in Medical Practice*. London: Duke University Press.

Morgan, D. L., & Kreuger, R. A. (1997). *The Focus Group Kit. Volumes 1–6*. Thousand Oaks, CA: SAGE.

Murdock, G. (1997). Thin descriptions: Questions of method in cultural analysis. In J. McGuigan (Ed.), *Cultural Methodologies* (pp. 178–192). London: SAGE.

Neyland, D. (2008). *Organizational Ethnography*. London: SAGE.

O'Donnell, E., Koch, B., & Boone, J. (2005). The influence of domain knowledge and task complexity on tax professionals' compliance recommendations. *Accounting, Organizations and Society*, *30*(2), 145–165.

Ochs, E. (1979). Transcription as theory. In E. Ochs & B. B. Schieffelin (Eds.), *Developmental Pragmatics*. New York: Academic Press.

Oreskes, N. (2019). *Why Trust Science?* Princeton: Princeton University Press.

Pettigrew, A. M. (1990). Longitudinal field research on change: Theory and practice. *Organization Science*, *1*(3), 267–292.

Pink, S., & Morgan, J. (2013). Short-term ethnography: Intense routes to knowing. *Symbolic Interaction*, *36*(3), 351–361.

Power, M. (1997). *The Audit Society. Rituals of Verification*. Oxford: Oxford University Press.

Rapley, T. (2004). Interviews. In C. Seale, G. Gobo, J. F. Gubrium, & D. Silverman (Eds.), *Qualitative Research Practice* (pp. 16–34). London: SAGE.

Rennstam, J., & Wästerfors, D. (2018). *Analyze! Crafting Your Data in Qualitative Research*. Lund: Studentlitteratur.

Richardson, L. (2000). Writing: A method of inquiry. In N. K. Denzin & Y. S. Lincoln (Eds.), *Handbook of Qualitative Research* (2nd ed.). Thousand Oaks, CA: SAGE.

Rienecker, L., & Jørgensen, P. S. (2018). *The Good Paper: A Handbook for Writing Papers in Higher Education* (2nd ed.). København: Samfundslitteratur.

Roethlisberger, F. J., & Dickson, W. J. (1939/1947). *Management and the Worker: An Account of the Research Program Conducted by the Western Electric Company, Hawthorne Works*. Chicago & Cambridge, MA: Harvard University Press.

Roulston, K. (2010). Considering quality in qualitative interviewing. *Qualitative Research*, *10*(2), 199–228.

Sandberg, J., & Alvesson, M. (2010). Ways of constructing research questions: Gap-spotting or problematization? *Organization*, *18*(1), 23–44.

Silverman, D. (1989). Six rules of qualitative research: A post-romantic argument. *Symbolic Interaction*, *12*(2), 215–230.

Silverman, D. (2000). *Doing Qualitative Research. A Practical Handbook*. London: SAGE.

Silverman, D. (2007). *A Very Short, Fairly Interesting and Reasonably Cheap Book about Qualitative Research*. London: SAGE.

Stake, R. E. (2000). Case studies. In N. K. Denzin & Y. S. Lincoln (Eds.), *Handbook of Qualitative Research* (2nd ed.). Thousand Oaks, CA: SAGE.

Starbuck, W. H. (2003). Turning lemons into lemonade: Where is the value in peer reviews? *Journal of Management Inquiry*, *12*(4), 344–351.

Strauss, A., & Corbin, J. (1990). *Basics of Qualitative Research: Grounded Theory Procedures and Techniques*. Newbury Park: SAGE.

Sveningsson, S., Alvehus, J., & Alvesson, M. (2012). Managerial leadership: Identities, processes, and interactions. In S. Tengblad (Ed.), *The Work of Managers* (pp. 69–86). Oxford: Oxford University Press.

Swales, J. M. (1990). *Genre Analysis. English in Academic and Research Settings*. Cambridge: Cambridge University Press.

Sword, H. (2012). *Stylish Academic Writing*. Cambridge, MA: Harvard University Press.

Van Maanen, J. (1979). The fact of fiction in organizational ethnography. *Administrative Science Quarterly, 24*(4), 539–550.

Van Maanen, J. (1988). *Tales of the Field. On Writing Ethnography*. Chicago: The University of Chicago Press.

Weber, M. (1914/1968). *Economy and Society. Vol. 1*. Berkeley: The University of California Press.

Weick, K. E. (1989). Theory construction as disciplined imagination. *Academy of Management Review, 14*(4), 516–531.

Wittgenstein, L. (1958). *Philosophical Investigations*. Oxford: Blackwell.

Yin, R. K. (1994). *Case Study Research*. Beverly Hills, CA: SAGE Publications.

INDEX

Pages in *italics* refer to figures.